emission detectors

emission detectors

alexander i bolozdynya

National Research Nuclear University "MEPhI", Russia

W World Scientific

NEW JERSEY · LONDON · SINGAPORE · BEIJING · SHANGHAI · HONG KONG · TAIPEI · CHENNAI

Published by

World Scientific Publishing Co. Pte. Ltd.

5 Toh Tuck Link, Singapore 596224

USA office: 27 Warren Street, Suite 401-402, Hackensack, NJ 07601

UK office: 57 Shelton Street, Covent Garden, London WC2H 9HE

British Library Cataloguing-in-Publication Data
A catalogue record for this book is available from the British Library.

EMISSION DETECTORS

ISBN-13 978-981-283-405-8
ISBN-10 981-283-405-2

Printed in Singapore.

To my *Alma Mater Studiorum*
Moscow Engineering Physics Institute

Foreword

This book was written by a researcher who has been working actively in a field of detector technology originally introduced almost 40 years ago by a group of Russian physicists investigating electron emission from liquid argon. They were able to formulate the unique properties of a new detector technology that could be developed employing this effect. They called these detectors *emission detectors*, meaning detectors of ionizing radiation, in which ionization electrons could be extracted from a condensed matter detection medium into a rarefied gas phase in which the electron signal could be amplified. Such detectors are sometimes called *two-phase* or *dual-phase detectors*, but that does not truly reflect the basic principle of their operation, therefore probably the name *two-phase emission detectors* would be more accurate.

Researchers working in emission detector technology are concerned with effects associated with quasi-free electron transmission through bulk dielectrics and charge carrier penetration through the inter-phase boundary, purification technologies that provide long electron lifetimes, methods of amplification of electron signals in pure dielectrics, analogue and digital visualisation of ionization particle tracks with extremely low ionization yield in condensed matter.

First considered for accelerator experiments hunting for rare processes, such as short-ranged particles, neutral channels of antiproton-nucleon annihilations in heavy nuclei and abnormal low-ionizing particles as well as medical imaging, emission detectors seem to have found their niche in low-background experiments searching for cold dark matter and low-energy neutrino interactions. The further development of this interesting class of radiation detectors may enhance the arsenal of modern experimental methods in other exciting fields like the neutrinoless double-beta decay search, solar neutrino physics and possibly in many new areas that cannot be satisfactorily served with existing detector technologies. In this sense, this monograph may be useful in providing readers with a solid foundation for development of the new instrumentation.

Boris A. Dolgoshein
Moscow
9 May 2009

Preface

This book is the first attempt to provide a comprehensive presentation of the physics and engineering concepts behind emission detector technology. Over the course of 35-year career in experimental physics, I have been involved in a variety of R&D projects that spans the study of effects associated with quasi-free electron emission, from small, gram-scale quantities of non-polar dielectrics for the purpose of exploring all possible working media for emission detectors, to the recent development of huge, ton-scale, astroparticle emission detectors used in the search for cold dark matter in the Milky Way. Consideration of this research itself can serve as a good example of the evolution of a new technology from basic concept to practical application.

In this book, special attention has been paid to the properties of non-polar dielectrics, since such media can be used for effective detection of penetrating radiation in many important practical applications, ranging from medical imaging to nuclear reactor diagnostics. Applications of emission detector technology are discussed in the concluding chapter of the book. More than two hundred references have been cited in this book and can serve as sources of more detailed information on particular subjects. For additional reading on relevant detector technologies, one can recommend *Liquid Ionization Detectors* (Atomizdat, 1983) by Barabash and Bolozdynya, *Noble Gas Detectors* (Wiley, 2006) by Aprile, Bolotnikov, Bolozdynya and Doke, and, of course, *Radiation Detection and Measurements* (3rd edition, Wiley, 2000) by Knoll, whose comments on the technology of noble gas detectors inspired me to work on this monograph.

I thank my colleagues from LUX collaboration, XENON-10 collaboration, CDMS collaboration, Moscow Engineering Physics

Institute, Institute for Theoretical and Experimental Physics, Case Western Reserve University, Massachusetts Institute of Technology, Purdue University, the Nuclear Medicine Group of Siemens Medical Systems, Department of Medical Physics of Rush-Presbyterian-St. Luke's Medical Center, Constellation Technology Corporation and 2K Corporation for many stimulating discussions on the issues related to the subject of the book. I also thank the many great scientists and nuclear engineers that I have had the privilege to work with on the development of noble gas and emission detectors, who have guided me to a better understanding of the physics and properties of these detectors and who helped in the routine of experimental activity. In particular, I am grateful to B.U. Rodionov, D. Akerib, D. Akimov, S. Anisimov, E. Aprile, K. Arisaka, R. Austin, A. Balakin, A. Barabash, A. Bernstein, A. Bolotnikov, A.F. Borghesani, P. Brusov, G. Carugno, W. Chang, M. Chen, V. Chepel, D. Churakov, C.E. Dahl, A. Dahm, R. DeVito, V.V. Dmitrenko, T. Doke, A.G. Dolgolenko, B.A. Dolgoshein, O.K. Egorov, V.V. Egorov, R. Gaitskell, K. Giboni, C. Hagmann, D. Koltick, R. Konoplich, J. Kwong, P. Lebedev, V.N. Lebedenko, M.I. Lopes, V.K. Lyapidevsky, K. Matthews, D. McKenzie, S. Medved, V.P. Miroshnochenko, V. Morgunov, I.M. Obodovsky, B.M. Ovchinnikov, S. Pokachalov, A.P.L. Policarpo, V. Prutskoj, A. Rostovtsev, B. Sadolet, G. Safronov, V. Shilov, T. Shutt, R.S. Shuvalov, E. Shuvalova, S.A. Simonychev, V.A. Smirnitsky, M. Smolin, G. Smirnov, V. Solovov, V. Sosnovtsev, P. Sorensen, V. Stekhanov, T. Sumner, Yu. Tikhonov, S. Ulin, L. de Viveiroz, H. Wang, J. White, B. Yakovlev and M. Yamashita.

I extend my special thanks to R. Austin, A. Bradley, M. Dragowsky, J. Kwong, C. Hall and P. Sorensen, who spent their time reading, and commenting on early versions of the manuscript, and A. Bernstein, E. Aprile, K. Arisaka, A.G. Dolgolenko, R. Gaitskell, B.U. Rodionov, T. Sumner, L. de Viveiroz, H. Wang and J. White, who contributed unique illustrations, and, especially, my family for their patience and moral support.

Alexander Bolozdynya

Contents

Chapter 1

Hetero-phase Detectors and History of Development of Emission Detectors

Properties of detectors used to detect, track and identify high-energy particles and radiation directly depend on the properties of their working media. Effective detection of low-ionizing and highly penetrative radiation is only possible with working media of sufficient mass and density. However, the amount of energy deposited by detected radiation is typically very small and special measures are needed to amplify signals generated by interactions of detected radiation with the detector's working media. There are several approaches for measuring the small amount of energy deposited in large mass detectors (see, for example, a classic monograph by Glenn F. Knoll). In this chapter, we focus our attention to a class of detectors utilising specific properties of hetero-phase working media in which detected radiation interacts with a condensed medium but signals are generated in the rarefied phase co-existing with the condensed medium or signals are generated in result of phase transitions.

1.1 Photo-emission detectors

The *photo-electric effect* is a quantum electronic phenomenon in which electrons are emitted from matter after the absorption of energy from electromagnetic radiation such as X-rays or visible light. The emitted electrons can be referred to as *photo-electrons* in this context. The effect was termed the *Hertz effect* after Heinrich Rudolf Hertz discovered in 1887 that UV radiation ignites spark between charged zinc electrodes. Further study of the effect led to understanding the quantum nature of light and to development of various detectors of photons and ionizing particles. In the months following Hertz's discovery, Alexander Stoletow investigated the photo-electron emission from different materials at different conditions (pressure and temperature) and found the electron emission from liquid solutes and Augusto Righi discovered photo-electron emission from solid dielectrics.

In 1902, P.E.A. von Lenard observed that the energy of the emitted electrons increased with the frequency, or colour, of the light. This was at odds with J.C. Maxwell's wave theory of light, which predicted that the energy would be proportional to the intensity of the radiation. The big surprise was that the photo-electric effect only occurs when a metallic surface is exposed to electromagnetic radiation above a certain threshold frequency. In 1905, Albert Einstein found a way to solve this paradox by describing light as composed of discrete quanta (photons) rather than continuous waves. Using M. Planck's theory of black-body radiation, Einstein theorised that the energy in each quantum of light was equal to the frequency multiplied by a constant, later called Planck's constant. A photon above a threshold frequency has the required energy to eject a single electron, creating the observed effect. This discovery led to the quantum revolution in physics and earned Einstein the Nobel Prize in 1921.

The discharge device of Hertz could be considered as the first emission detector sensitive to UV radiation. Since the discovery in the late 1920s that electrons accelerated in vacuum may cause emission of secondary electrons and as a result provide amplification of the electron signal, photo-multipliers begin their distinguished history. Photo-electron multiplier tubes (PMT) based on the multi-stage amplification of a photo-electron signal was invented by L.A. Kubetsky in 1930 and re-invented by V. Zworykin in 1936. Since that time photo-emission detectors received a long and well-proven record in experimental physics, nuclear medicine, oil well logging, monitoring nuclear materials and many other important technical applications.

1.2 Cloud chambers

At the boundary between the 19th and 20th centuries, Scottish physicist Charles T.R. Wilson, studying cloud formation and optical phenomena in moist air, discovered that ions could act as centres for water droplet formation (see, for example, excellent review by Henderson, 1970). Wilson built a chamber filled with saturated vapour of water-alcohol mixture and connected it to an expansion mechanism. By rapidly reducing the pressure, the gas inside the chamber undergoes an adiabatic expansion and cools; supersaturated vapour condenses in liquid droplets on centres of nucleation, which could be dust or ions. The condensation of liquid along tracks of ionizing particles makes them visible. Many physicists experimented with such devices. In 1924, Peter L. Kapitza placed the *Wilson chamber* in a strong magnetic field that allowed measurements not only the charge but also the momentum of particles analysing curvature of their tracks. The new approach led Carl D. Anderson to discovery of the positron in 1932 (winning him the Nobel Prize in Physics for 1936) and the muon in 1937 in cosmic rays.

Wilson, along with A.H. Compton, received the Nobel Prize for Physics in 1927 for his work on the cloud chamber. This kind of chamber is also called a *pulsed chamber*, because the conditions for operation are not continuously maintained.

In fact, the Wilson chamber was the first device using a two-phase (saturated vapour and liquid droplets) working medium for detection and visualisation of tracks of individual particles. Since then, diffusion, spark and streamer cameras have been developed for visualisation of particle tracks in gases at normal pressure. With the ever increasing energies of investigated particles, condensed working media have been considered for construction of more sensitive devices.

1.3 Bubble chambers

A *bubble chamber* is an extension of the idea of cloud chamber to denser working media (see, for example, a great review by Aleksandrov *et al.*, 1967). The bubble chamber was invented by Donald Glaser in 1952, when he has shown that ionizing radiation could initiate boiling in superheated liquids. Soon after that he succeeded in photographing a string of vapour bubbles induced by the passage of an ionizing particle through a superheated liquid. The dramatic difference in optical density between liquid and vapour was used to photograph tracks of high-energy particles interacting with liquids. The bubble chamber become the second example of a detector using a two-phase (liquid and saturated vapour in bubbles) working medium. In 1960, Glaser received a Noble Prize for the development of bubble chambers.

The first physical results with bubble chambers were obtained with 'clean' bubble chambers made of glass. One reason for this approach is that the surface of the glass is free from microscopic defects that may become centres for the initiation of boiling of the liquid. In addition, glass chambers may be made without seams or seals. However, with technological improvements, tracks were soon successfully observed in 'dirty' chambers made of a polished metal body and flat glass windows sealed with Teflon. Eventually, most of the cryogenic liquids found application in bubble chambers, with helium and xenon among them (Table 1.1). Light liquids allow using magnetic fields in order to determine the momentum of charged particles from the curvature of their tracks. The heavy liquids are attractive because of their efficient absorption of gamma radiation. Since heavy noble liquids are the most popular working media of modern two-phase emission detectors, we refer to liquid xenon bubble chambers as a direct prototype of the emission detectors.

Table 1.1 Properties of bubble chamber liquids (Harigel, 2003; Henderson, 1970).

Liquid	Temperature, K	Vapour pressure, bar	Density, g/cm^3	Radiation Length X_0, cm	Absorption Length λ_0, cm
He	3.2	0.4	0.14	1027	437
H_2	25	4	0.0645	968	887
D_2	30	4.5	0.14	900	403
C_3H_8	333	21	0.43	110	176
Ne	35	7.7	1.02	27	89
Ar	135	28	1.0	20	84
CF_3Br	303	21	1.5	11	73
Xe	253	26	2.2	3.9	56

Xenon, being predominantly a monatomic medium, has no rotational or vibration atomic oscillation modes, and as a result, it is effectively converting the energy of δ-electrons into light (scintillation). To convert the energy of scintillation into localised heat and enhance the formation of bubbles, molecular admixtures of ethylene or propane have been used in LXe bubble chambers (see references in Table 1.2)

Table 1.2 Liquid xenon bubble chambers.

Built	Temperature, K	Density, g/cm^3	Working Volume	Reference
1950s	254	2.3	Ø2.5·1.3 cm^2	Brown *et al.*, 1956
1950s	254	2.2	Ø30·25 cm^2	Brown *et al.*, 1961
1960s	260	1.4	180 litres	Barmin *et al.*, 1972
1970s	260	1.4	1.5·0.7·0.7 m^3	Barmin *et al.*, 2003

The first liquid xenon bubble chamber was built by Glaser in the 1950s, the largest and the most successful *DIANA LXe bubble chamber* was built at ITEP (Moscow) in the 1970s (Fig. 1.1). Images acquired from LXe bubble chambers allow for the reconstruction of topologically complicated events, including large number of γ-quanta and π^0-mesons. DIANA was decommissioned in 1985 but a data analysis from about 1 million stored images is still continuing. Using this archive, ITEP scientists recently reported the discovery of exotic *pentaquark* resonance generated in low-energy K$^+$Xe collisions (Barmin *et al.*, 2003).

Figure 1.1 Liquid xenon bubble chamber DIANA with 1.5x0.7x 0.7 m^3 active volume constructed at ITEP in the 1970s. Courtesy of A.G. Dolgolenko.

In an attempt to improve the performance of bubble chambers and integrate them into multi-detector systems with sophisticated triggers for high energy experiments, the use of the scintillation properties of some liquids was considered. One of the first bubble chambers of this type was a three-litre chamber filled with liquid propane to which a naphtalene wave-shifter was added (Minehart and Milburn, 1960). From this work, it was shown for the first time that scintillation and bubble nucleation are not necessarily mutually exclusive. Later, the use of a scintillation trigger and the collection of charge during the expansion cycle in a liquid argon bubble chamber were considered by Berset *et al.* in 1982. The amplitude of the scintillation trigger can give preliminary information about the energy deposited by electromagnetic or hadronic showers; the scintillation signal can be used to trigger a laser beam to initiate marker tracks simultaneously with useful events.

These first attempts demonstrated that bubble chamber technology would not be easily integrated into fast detectors with an electronic readout. That was the original motivation to explore new approaches in detector technology based on condensed working media and electronic *filmless imaging*.

At the beginning of the 21st century, bubble chambers again attracted the attention of particle physicists, who, at this time, were interested in searching for cold dark matter in the form of weakly ionizing massive particles (WIMPs; Bond *et al.*, 2005). At the University of Chicago, it was shown that, for certain pressure and temperature operating conditions, the vaporisation of the overheated liquid can only be produced by particles having a large stopping power, such as nuclear recoils, making the detector insensitive to most minimum ionizing backgrounds due to gamma radiation (Collar *et al.*, 2000). The devices are operated at near room temperature and use non-flammable, non-toxic and inexpensive liquids, with a chemical composition (e.g. CF_3I) that provides maximum sensitivity to neutralino interactions through both the spin-dependent and spin-independent channels.

1.4 Electron emission detectors

The development of emission detectors has a long history. The first emission detector was probably Hertz's spark gap triggered with ultraviolet radiation (1887). The discovery of photo-electron emission from metals eventually led to an understanding of the quantum nature of light and the development of the quantum age in modern physics. Since then, the effect has been successfully used in vacuum photo-diodes and photomultipliers to detect low-intensity photon fluxes or even single photons for a variety of technical applications. The discovery of electron emission from dielectrics by Stoletow and Righi extended the technology for massive detection media and made it possible to detect extremely low-ionizing radiation and very rare particle interactions.

The next important step was made only half a century later, in 1948, after the discovery (Davidson and Larsh, 1948) that ionization electrons can live long enough in condensed argon to be collected from the bulk material and their charge to be measured. The first electron emission two-phase (solid and gas argon) detector was constructed by a graduate student, G.W. Hutchinson, in 1948 (Hutchinson, 1948). However, the advantages of the new approach were not fully recognised at that time by the detector science community, and the idea was forgotten for about 20 years.

The advantages of condensed noble gases for precision imaging and for the development of detectors for high energy particles were recognised by L. Alvarez in 1968. The development of liquid xenon ionization chambers by L. Alvarez, H. Zaklad, S. Derenzo at Lawrence Berkley National Laboratory, in the 1960s and 1970s, seemed to open a way for digital imaging 140–511 keV gamma rays for nuclear medicine needs with noble gas detectors.

Independently of Alvarez, the possibility of using condensed noble gases as working media for particle detectors attracted the attention of B.A. Dolgoshein and his co-workers at Moscow Engineering Physics Institute. They were led by B.U. Rodionov, who rediscovered electron emission from liquid argon and observed electroluminescence of pure noble gases in the course of attempting to develop a high-density *streamer chamber*. They proposed using the properties of noble gases to develop highly sensitive imaging instrumentation with electronic readout — *emission detectors* (Rodionov, 1969; Dolgoshein, Lebedenko and Rodionov, 1970). Since then, new emission detector media have been discovered and a variety of different emission detectors has been developed as described in this book. In general, it was shown that emission detectors can combine the high detection efficiency inherent to dense and massive working media with high gain through the physical amplification of electron signals that is only possible in the rarefied phases (Rodionov, 1969, 1987).

The first attempts to construct emission detectors were focussed on developing imaging cameras to visualise tracks of individual relativistic particles in high energy physics experiments (Bolozdynya *et al.*, 1977). However, soon after a successful test of the first emission streamer chamber at the ITEP proton synchrotron, it was understood that the long drift time required to operate emission cameras drastically limited their utility in experiments in modern high-luminosity accelerators.

After that, the emission method was explored for imaging gamma radiation for nuclear medicine applications (Egorov *et al.*, 1983). The first attempt to develop such a gamma camera was quite successful and a group of young physicists (S. Kalashnikov, V. Egorov and A. Bolozdynya), working on the development of an emission electroluminescence gamma camera, received an award from the USSR Academy of Science in 1984.

With the development of *non-accelerator physics* and *low-background experiments* searching for rare nuclear decays and exotic particles, it became clear that, in this field, emission detectors can demonstrate all their advantages, including the possibility of constructing 'wall-less' detectors with large mass and low detection threshold (Bolozdynya *et al.*, 1995). In the relatively short period that has passed since that time, emission detectors have filled a unique niche in the arsenal of modern instrumentation for experimental and applied physics. It is quite notable, that for the last thirty years, emission detector technology has developed from the miniature R&D devices of about $1\,cm^3$ sensitive volumes to multi-tonne detectors, as described in Chapter 8.

Chapter 2

Emission of Charge Carriers from Working Media of Emission Detectors

Since emission detectors employ effective emission of charge carriers from condensed media, we must consider conditions for effective electron and ion emission from metals, semiconductors and dielectrics that can be used for those purposes. In particular, attention should be devoted to non-polar dielectrics, such as condensed noble gases and liquid saturated hydrocarbons that can better perform in the role of bulk and massive working media of emission detectors. As shown in the following chapters, such detectors have found the most practical applications in detection of exotic low-ionizing particles and imaging of radiation fields of high-energy radiation.

2.1 Electron emission from metals

The *free-electron theory* of metals introduced by Sommerfeld in 1938 has provided the basis for practically all quantum-mechanical theories of electron emission from metals until recently (Modinos, 1984). We shall use some important formulas derived on basis of this theory for consideration of experimental data related to working media of emission detectors.

In the free-electron theory of metals, the electron states in the conduction band are described as standing plane waves in a resonator size of the metal. Bearing in a mind that there are two electron states with different spin orientation for the same spatial wave function, the probability of an electron state with energy E being occupied is given by the *Fermi-Dirac distribution function*

$$f(E) = \frac{1}{1 + \exp[(E - E_F)/k_B T]} \qquad (2.1)$$

where k_B is the Boltzmann's constant, T is the absolute temperature, and E_F is the Fermi level. The latter is the highest occupied quantum state in a system of fermions at absolute zero temperature and for electrons in metal given by

$$E_F = \frac{\hbar^2}{2m}(3\pi^2 N_c)^{2/3} \qquad (2.2)$$

where m is the electron mass, N_c is the number of conduction band electrons per unit volume that is an empirical parameter.

2.1.1 Field emission

Let us consider a semi-infinite metal occupying the half-space from $z = -\infty$ to $z = 0$ and neighbouring with vacuum at $z > 0$. An electron situated at a finite distance in the vacuum is attracted to the metal surface by an *image force* $-e^2/4z^2$. The potential energy of the electron on the vacuum side of the metal-vacuum interface is asymptotically given by

$$V(z) \cong E_F + \varphi - e^2/4z^2 \qquad (2.3)$$

where φ is the work function. Equation 2.3 is considered to be valid for $z > 3$ Å (Modinos, 1984). Formally, in free-electron theories it is assumed that Eq. 2.3 remains valid up to the point z_c such that $V(z_c) = 0$ (see Fig. 2.1). When an external electric field F is applied to the surface, the potential energy is changed by adding the term $-eFz$. Since in free-electron theories it is assumed that the electric field can not penetrate inside the metal, in electron emission experiments ($F > 0$) the potential energy near the metal surface is described by the following

$$V(z) = E_F + \varphi - e^2/4z^2 - eFz, \ z > z_c$$
$$V(z) = 0, \ z \le z_c \qquad (2.4)$$

The potential forms a barrier for escaping electrons at the surface of metal as shown in Fig. 2.1. Note that the peak of the barrier occurs at

$$z_m = (e/4F)^{1/2} \tag{2.5}$$

and at this point it has a value

$$V_{max} = E_F + \varphi - (e^3 F)^{1/2} \tag{2.6}$$

One can see that the barrier height is reduced by the value $(e^3 F)^{1/2}$. This barrier height reduction is called the *Schottky effect*.

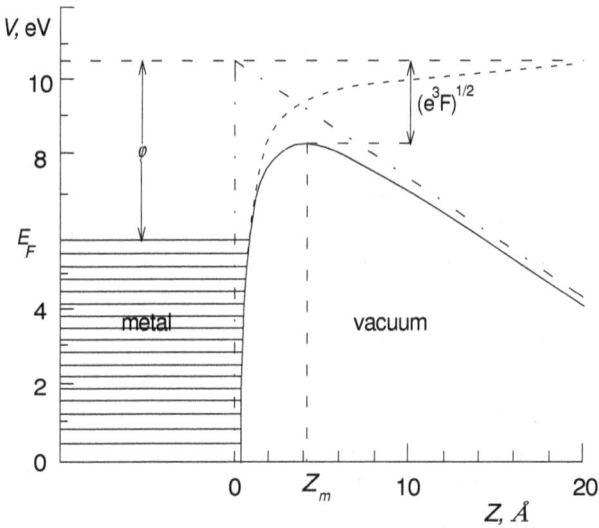

Figure 2.1 The surface potential barrier seen by an electron in a field emission experiment from metal. The contribution of the image potential and the applied field are shown by the broken and the broken solid line, respectively. Redrawn from Modinos, 1984.

An electron can penetrate the barrier with *transmission probability* $D(E_z)$ which depends on its momentum in the normal direction p_z or normal energy

$$E_z = p_z^2 / 2m + V(z) \tag{2.7}$$

If the number of electrons per unit area per second $N(E_z)$ with the z-component of their energy within dE_z of E_z incident on the surface potential barrier, the electron emission current density is given by integral

$$J = e \int_0^\infty D(E_z) N(E_z) dE_z \qquad (2.8)$$

At relatively low temperatures when thermoelectric emission (see Section 2.1.2) could be neglected, the tunnelling current follows the *Fowler–Nordheim equation*

$$J(F) = AF^2 \exp[-B\varphi^{3/2} / F] \qquad (2.9)$$

and the field emission is called *Fowler–Nordheim tunnelling*.

The model assumes that the electrons in the metal remain at equilibrium, despite the electrons escaping the metal surface. *A* and *B* are weak functions of *F* and in the narrow region of field strength of a typical field emission experiment ($0.3 \leq F \leq 0.5$ V/Å or MV/cm) can be considered as constants. Then, plotting the current density as $\log(J / F^2)$ versus $1/F$ makes the corresponding curve look almost like a straight line. This representation of the field emission experiments has been widely used for determination of work functions from the slope of the plot. The linearity of the Fowler–Nordheim plot served as evidence of the thermoelectric nature of electron emission.

2.1.2 Thermionic emission

At high temperatures and relatively weak applied electric fields, charge carriers can effectively penetrate the surface potential barrier due to energy of the medium. The most classical example of thermionic emission is the emission of electrons from a hot metal cathode into a vacuum, historically known as the *Edison effect* but first discovered by D. Lordan in 1873. The tunnelling current increases dramatically with increasing temperature and for vacuum emission from metals it tends to only become significant for temperatures over 1000 K.

In the weak applied field approximation, the tunnelling current Eq. 2.8 reduces to the Schottky formula

$$J(F,T) = A_R T^2 \exp\left[-\frac{\varphi - (e^3 F)^{1/2}}{k_B T} \right] \qquad (2.10)$$

where the proportionality constant, given by

$$A_R = \frac{emk_B^2}{2\pi^2 \hbar^3} = 1.20173 \cdot 10^6 \, \text{Am}^{-2}\text{K}^{-2} \qquad (2.11)$$

is known as the Richardson's constant. Putting $F = 0$ in Eq. 2.10, one can obtain the well-known Richardson's (or *Richardson–Laue–Dushman's or RLD*) equation

$$J(T) \approx A_R T^2 \exp(-\varphi / k_B T) \qquad (2.12)$$

The equation was originally derived from experimental data by Richardson (1902, 1912), and independently by Von Laue (1918), prior to the discovery of quantum mechanics. In 1923, Dushman gave the first quantum mechanical explanation of the thermionic emission. The complete theory was developed by Fowler and Nordheim (1928) and Nordheim (1928) after discovery of the spin of electron.

Probing electron emission with the RLD-equation has been used to prove that the 'fast' electron emission from noble liquids, in fact, is the thermal emission of electrons heated by the electric field (see below Section 2.3.1).

2.1.3 Photo-electron emission

When a medium containing electrons is exposed to electromagnetic radiation above a certain threshold frequency, electrons acquire energy large enough to be emitted. Historically, an electron emission stimulated by the absorption of light was the first experiment demonstrating an effect of electron emission from condensed media. The *photo-electron emission* was discovered by Hertz in 1887 and the effect now bears his name his name. Hertz found that spark discharge between charged electrodes may be initiated with ultra-violet (UV) illumination of the electrodes, especially, negative. Hertz experimented with two discharge gaps using one of them as a source of the UV radiation and another one

for measurement of the minimal gap when the discharge occurred for a
given voltage (Fig. 2.2). The primary circuits of two Ruhmkorff coils *a*
and *e* were placed in series with a battery *b* and interrupter *c*. The first
coil was provided with a fixed gap *d* in the secondary circuits, the second
with a variable, micrometer, gap *f*. Hertz found that longer sparks could
be obtained across *f* when it was exposed to the spark across *d*. Using
different optical filters *p*, he demonstrated that the effect is associated
with UV irradiation of clean metal electrodes.

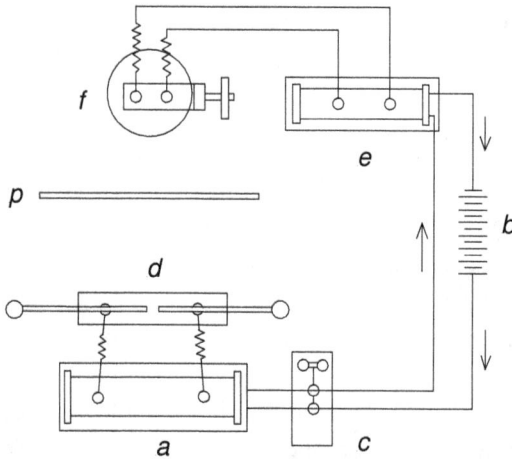

Figure 2.2 Schematic drawing of the Hertz experiment on discovery of
photo-electron emission from metal in 1887: *a* and *e* are the primary
circuits of two Ruhmkorff coils, *b* is the battery, *c* is the interrupter, *d* is
the fixed discharge gap in the secondary circuits, *f* is the discharge gap
variable with micrometer. Redrawn from Zworykin and Ramberg, 1949.

Shortly afterwards, Stoletow (1890) increased the sensitivity of the
photo-electron emission experiments by putting a voltaic battery in series
with the electrode gap. He also observed for the first time a photo-
electron emission from liquids. Elster and Geitel (1890) obtained
daylight sensitive photo-cathodes by preparing them from alkali metals
deposited in vacuum. This was the first true photo-tube.

After series of experiments, it was understood that the effect is
associated with the emission of electrons from metals under the influence
of light at a certain frequency ν and the kinetic energy of electrons E_e is

reduced for the *work function* of metal

$$E_e = \hbar v - \varphi \qquad (2.13)$$

In 1905, the threshold behaviour of the photo-electron emission was explained by Einstein, who suggested that light is composed of discrete quanta, rather than continuous waves. This great idea led to the development of quantum mechanics, revolutionizing science, and to the Nobel Prize in Physics for its author in 1921.

The photo-electron emission process could be broken down into three consecutive events:

1) the absorption of photons in the photo-cathode;

2) the transport of the excited electron to the surface during which the electron is losing its original energy;

3) the escape of the electron across the potential barrier from the photo-cathode.

Exploration of the photo-electron effect led to the development of a family of extraordinary devices, such as photo-multipliers, widely used for the detection of single photons and imaging devices for television, revolutionizing human society (see classic review by Zworykin and Ramberg, 1949).

2.1.4 Electron emission into gas

Most of the experiments of electron emission from metals have been performed in vacuum. However, for radiation detectors the process of electron emission into the rarefied gas phase is of special interest because it can be used for physical amplification of the electronic signal. In 1890, Stoletow discovered that there is an essential difference in the emission of electron between vacuum and gas. Being emitted in gas electrons interact with atoms of gas and this interaction may lead to the following effects:

1) backscattering electrons into the emitter;

2) ionization of the gas that leads to building space charge of low-mobility ions around the emitter;

3) the capture of electrons by electronegative atoms that affects the built-up space charge in vicinity of the emitter;

4) the drift of electrons away from the emitter;

5) excitation of the gas during the drift of electrons that may lead to emission of photons and secondary photo-electron emission eventually developing into discharge.

In fact, in the first experiments performed by Hertz with spark gaps, the field and photo-emission of electrons from metal cathodes into gas played the major role.

At earlier stages of the development of photo-multipliers, when charge-sensitive electronic amplifiers were not yet available, the gas-filled photo-tubes were considered for detection of weak light signals (Zworykin and Ramberg, 1949). At that time, the Radio Corporation of America commercially produced the gas-filled photo-tubes with a *gas gain* (electron multiplication factor) of about 10 and sensitivity of about 10–100 nA/μW in the visible and near-infrared range.

2.2 Electron emission from semiconductors

In contrast to the situation with metal emitters, in semiconductors electron emission even at relatively small fields may change the equilibrium distribution of electrons because of deep penetration of the applied electric field into the semiconductor. In formal language, the field penetration may be described as the band structure bending in the vicinity of the surface. Field emission from semiconductor surface can be due to electron tunnelling from the conduction band, valence band and sometimes from the surface states as shown in Fig. 2.3.

The most practical applications have been found for semiconductive photo-cathodes with *negative electron affinity* of the surface (NEA), when the bottom of the conduction band lies above the potential barrier at the surface (Fig. 2.3b). In practice, the condition may be obtained by heavy p-doping (of order 10^{19} cm^{-3}) of the semiconductor (to encourage downward band bending at the surface) and by adding a thin film (several atomic layers) of cesium-enriched cesium oxide (Spicer, 1977).

The threshold of response of a NEA photo-cathode is set by the semiconductor band gap. The NEA photo-cathodes are often used as materials for PMT dynodes with high secondary electron emission yield (typically 30 at 400 eV for GaP/Cs). This is a result of effective escape of electrons from the depth of about the diffusion length which can be as large as 10 μm. In other words, the electrons can be thermalised into the conduction band minimum then diffuse to the surface and still effectively

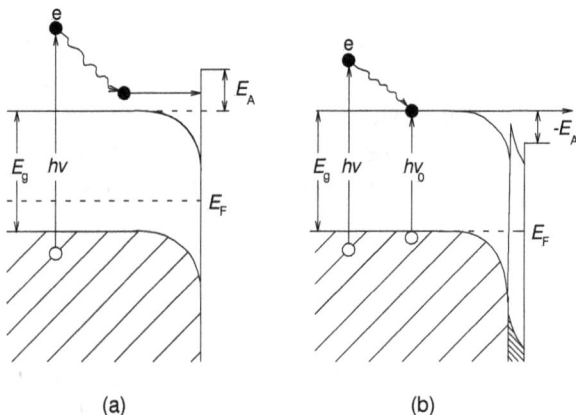

(a)   (b)

Figure 2.3 Emission of electrons from semiconductor with positive (a) and negative (b) affinity of the surface to electrons. Redrawn from Modinos, 1984.

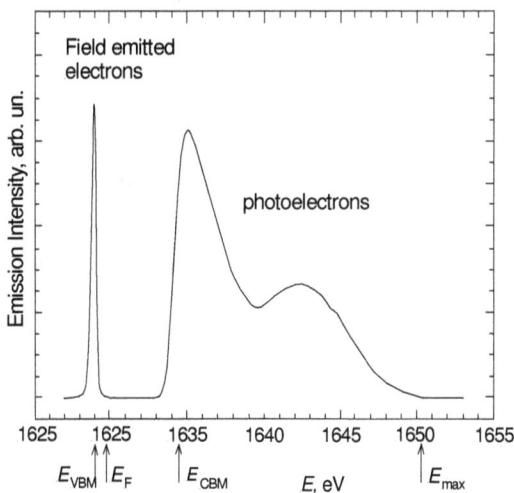

Figure 2.4 Spectrum of electrons emitted from diamond biased at −1630V under influence of 21.2 eV laser radiation and electric field. Redrawn from Bandis and Pate, 1996.

escape. In NEA photo-cathodes, a 'built-in' field associated with the band bending at the surface accelerates electrons, allowing them to penetrate the surface barrier effectively. The thermoionic emission from NEA photo-cathodes provides the fundamental noise especially for the lower band gap semiconductors. Such devices are often cooled down for this reason. Due to the relatively large escape depth for electrons, the devices found application for the detection of infrared radiation, for example, in night-vision devices.

In real experiments, the field emission and the photo-electron emission are often present at the same time. For example, Fig. 2.4 shows the electron emission spectrum measured with a sample of diamond biased at $V = -1630$ V and illuminated with 21.2 eV laser radiation. The energy spectrum consists of two peaks: one due to field-emitted electrons ($E_e < 1630$ eV) and the other due to photo-electrons ($E_e > 1630$ eV). Electrons photo-excited from the valence band are emitted with the highest energy $E_{VBM} - 21.2$ eV, where E_{VBM} is the valence band maximum occupation energy. The energy of photo-electrons emitted from the conduction band $E_{CBM} = -21.2\text{eV} + E_g$, where E_{CBM} is the minimum occupation energy of the conduction band. In this experiment, the field component of the electron emission is present independently of the illumination.

2.3 Electron emission from dielectrics

Since the dielectric constant of condensed dielectrics is different from that of vacuum or gas, charge carriers experience an *image potential A* near the interface acting against transmission of the charge into the less dense phase:

$$A_{1,2} = -e(\varepsilon_1 - \varepsilon_2) / \left[4\varepsilon_{1,2}(z + \xi z / |z|)(\varepsilon_1 + \varepsilon_2) \right] \qquad (2.14)$$

where $\varepsilon_1 F_1 = \varepsilon_2 F_2$, ε is the dielectric constant and Indexes 1 and 2 are related to the condensed phase and the equilibrium gas (vacuum) phase, respectively; z is a Cartesian coordinate perpendicular to the interface, ξ is the cutting parameter of about thickness of the inter-phase transition layer (a few nanometres or between 2 and 3 atomic monolayer according to Croxton, 1974). The image potential is temperature dependent, approaching zero at temperatures close to the critical temperature.

Charge carriers driven to the interface by an applied electric field $\vec{F}$ may be localised under the interface (if they can not penetrate the surface barrier of the image potential immediately) on the depth

$$z_0 = -\left[\frac{e}{4F_1} \frac{\varepsilon_1 - \varepsilon_2}{\varepsilon_1(\varepsilon_1 + \varepsilon_2)}\right]^{1/2} \tag{2.15}$$

If charge carriers are generated by ionizing radiation as a point-like cloud, being localised under the interface, they populate the volume, the linear size of which can be estimated as

$$\ell = 3k_B T / 2eF_1 \tag{2.16}$$

At favourable conditions as described in the following sections, some charge carriers can be extracted from condensed dielectrics through the free interface surface. In general, emission of charge carriers may be characterised with a specific *emission time* t_e needed to penetrate the interface barrier by some large number of electrons with an average velocity v_z in the normal direction to the interface:

$$t_e = \ell / v_z \tag{2.17}$$

The average velocity of charge carriers escaping through the interface can be calculated as

$$v_z = P \int_{p_z > p_0} \alpha\beta\frac{p_z}{m} f(\vec{p}) d\vec{p} \tag{2.18}$$

where P is the probability of the quasi-free electron state, α is the coefficient of penetration of the potential barrier depending on the barrier shape and wave-function of the charge carrier, β is the probability to avoid backscatter from gas, p_0 is the emission threshold momentum, $f(\vec{p})$ is the *momentum distribution function*.

For massive ions, the velocity of barrier penetration is extremely low and, in practice, ions may be only extracted from dielectrics due to instability of the liquid interfaces as considered in the Section 2.5. Electrons due to low effective mass can penetrate the interface potential barrier at realistic conditions in non-polar dielectrics. The probability to avoid backscatter from gas β can be calculated as

$$\beta = \frac{4v_2}{\bar{v} + 4v_2} \qquad (2.19)$$

where v_2 is the electron drift velocity in the gas, $\bar{v}$ is the average velocity of electrons escaped into the gas (Loeb, 1955).

For average extraction time t_e much greater than the relaxation time of the momentum distribution function $f(\vec{p})$, the emission can be considered as a stationary process, then, the number of charge carriers under the surface is described as

$$N(t) = N_0 \exp(-t/t_e) \qquad (2.20)$$

If the *lifetime of quasi-free electrons* is essentially limited by the electron capture time t_c, Eq. 2.19 should be corrected for this effect

$$N(t) = N_0 \exp(-t/t_e - t/t_c) \qquad (2.21)$$

and the total number of emitted electrons within a time t can be obtained by integrating the emission rate $dN(t)/dt$ that bring to the following result

$$N_e(t) = N_0 t_c (t_e + t_c)^{-1} \{1 - \exp[-(t_e + t_c)t/t_e t_c]\} \qquad (2.22)$$

More formally, the electron emission may be described in terms of the *probability (coefficient) of emission*

$$K_e = N_e(\infty)/N_0 = (1 + t_e/t_c)^{-1} \qquad (2.23)$$

As seen, the total number of all emitted thermal electrons is controlled by the lifetime of excess electrons before their attachment to electronegative impurities or capture into other deep traps.

2.3.1 Electron states in non-polar dielectrics

In the non-polar wide-gap dielectrics, there are no free electrons in the conduction band at normal state and the applied electric field easily penetrates bulk samples. Being excited with radiation, electrons from the valence band may escape into the conduction band and live there long enough time to be transported for long distances. In non-polar dielectrics, the electron life time is limited by the capture on electronegative impurities or structure traps in solids.

Electrons behave differently in different non-polar dielectrics. Obviously, their behaviour is classified via the value of mobility μ_0 in zero limit of the electric field. There are three classes of electron states classified via the *mobility in zero-field approximation*:

1) $\mu_0 > 10 \text{ cm}^2\text{V}^{-1}\text{s}^{-1}$,
2) $\mu_0 < 0.1 \text{ cm}^2\text{V}^{-1}\text{s}^{-1}$,
3) $0.1 \text{ cm}^2\text{V}^{-1}\text{s}^{-1} < \mu_0 < 10 \text{ cm}^2\text{V}^{-1}\text{s}^{-1}$.

Electrons with mobility $\mu_0 > 10 \text{ cm}^2\text{V}^{-1}\text{s}^{-1}$ are usually classified as *quasi-free electrons*. This term suggests that *free* electrons exist only in vacuum; however, in some condensed noble gases electrons are very mobile and demonstrate μ_0 comparable to that in low density gases. The highest electron mobility has been observed at the 'zero-field' approximation of the dependence of the electron drift velocity versus applied electric field in heavy condensed noble media such as solid xenon $(4500 \text{ cm}^2\text{V}^{-1}\text{s}^{-1})$ and isomers of saturated hydrocarbons with spherical molecules such as methane $(400 \text{ cm}^2\text{V}^{-1}\text{s}^{-1})$. However, the highest absolute values of electron drift velocities (up to 10^7 cm/s) are possible in liquid saturated hydrocarbons at electric fields (up to 100 kV/cm) since the dependence of the electron drift velocity versus applied electric field $v_{dr}(F)$ has no saturation in such media as it happens in condensed noble gases at electric fields $F \sim 1$ kV/cm.

Electrons with mobility $\mu_0 < 0.1 \text{ cm}^2\text{V}^{-1}\text{s}^{-1}$ are considered as being *localised* in deep traps or density fluctuations which develop into deep traps. In liquid light noble media, the atoms have a small polarisability (helium, neon, hydrogen) and electrons are localised in vacuum bubbles with a radius of $R \sim 1$ nm and mobility down to $10^{-3} \text{ cm}^2\text{V}^{-1}\text{s}^{-1}$ (see, for example, Khrapak, Schmidt and Illenberger, 2005). In normal liquids, the drifting electron bubble experiences hydrodynamic resistance and the mobility can be calculated in accordance with *Stokes's law*

$$\mu = \frac{e}{6\pi\eta R} \qquad (2.24)$$

where η is the *viscosity* of the liquid. In superfluid helium, mobility of localised electrons is essentially increased but still remains smaller than that of positive ions.

The third case is a transition class where localised and quasi-free states are possible at different probabilities. This situation is most clearly realised in liquid *n*-alkanes. In hydrocarbons, electrons are temporarily trapped in structural voids. Through thermal activation, electrons can be liberated from the traps and behave as free charge carriers until the next trapping. The resident time in a trap depends on temperature T and can be characterised with activation energy E_a specific for different liquids

$$t_{tr} = t_0 \exp\left(-\frac{E_a}{k_B T}\right) \qquad (2.25)$$

If free electron lifetime t_F, *mobility of free electron* μ_F and the trapped electron is immobile (in fact, it can drift as ion) in comparison with a free electron, the *quasi-free electron mobility* can be expressed as

$$\mu = \mu_F(T)\left(\frac{t_F}{t_{tr}}\right)\exp\left(-\frac{E_a}{k_B T}\right) \qquad (2.26)$$

The probability of electrons to exist in the *quasi-free state* can be calculated as a ratio of the ion mobility to the quasi-free electron mobility (150 ± 50 cm V^{-1}s^{-1} for liquid iso-octane at room temperature)

$$P = \frac{\mu}{\mu_0}. \qquad (2.27)$$

More details about behaviour of electrons in non-polar dielectrics may be found in monographs written by Schmidt (1997) and Barabash and Bolozdynya (1993).

In dense media, an electron interacts with a few atoms simultaneously. Atoms, polarised by the electron electric field, are attracted to the electron. At the same time, the polarised atoms interact with each other as dipoles. A balance between these interactions defines a potential *energy of the ground state of electrons* V_0 in the condensed dielectrics. Experimentally, V_0 is most often measured as a difference between work functions in two separate photo-electron emission experiments. In one experiment, the photo-cathode is placed in vacuum (or in very low density gas). In another one, the photo-cathode is placed in the tested condensed medium.

There is a clear correlation between V_0 and the electron mobility in the zero-field approximation μ_0: the mobility μ_0 reduces with increasing V_0 (Fig. 2.5). From Table 2.1 one can see that in liquid noble gases, changing V_0 from -0.5 eV to $+0.5$ eV leads to μ_0 changing in several orders of magnitude. In noble solids, μ_0 depends on V_0 almost linearly.

At the electric field $\vec{F}$, applied to electrons extracted from the condensed dielectric, and the energy of the ground state of quasi-free electrons V_0, the total potential energy of electrons in the vicinity of the

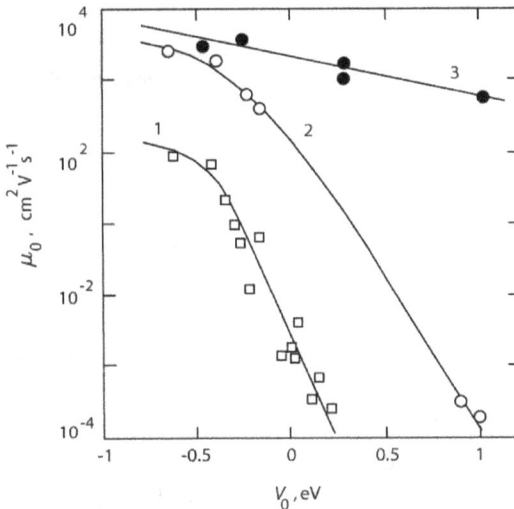

Figure 2.5 Correlation between zero-field electron mobilities and the ground state of electrons in liquid hydrocarbons (1), noble liquids (2) and noble solids (3) (Barabash and Bolozdynya, 1993).

Table 2.1 Electronic properties of liquid (L) and solid (S) non-polar dielectrics used in emission detectors (Bruschi *et al.*, 1975; Grand and Bernas, 1977; Barabash and Bolozdynya, 1993; Bolozdynya, 1999).

	T, K	ε	μ_o, cm²/V/s	V_o, eV	F_c, kV/cm	F_o, kV/cm	t_e
Emitters of cold electrons							
L⁴He	1-2	1.05	0.03	+1			10 s (100 V/cm)
Ln-H	300	1.88	0.09 (E_a = 0.19 eV)	+0.09	100	0.03	
Sn-H	77			+0.98			
Liso-O	300	1.94	7 (E_a = 0.05 eV)	−0.18	90	0.15	20 μs (1 kV/cm)
Siso-O	77			+0.57			
LTMP	297		24 (E_a = 0.06 eV)	−0.43	50		
STMP	77			+0.34			
LAr	84	1.51	475	−0.21	0.2		700 μs (100 V/cm)
LNe	35.7	1.18	0.001	+0.67		1.0	20 s (1 kV/cm)
Emitters of hot electrons							
SNe	24		600	+1.1			
LCH₄	100	1.66	400	−0.18	1.5	<4	
SCH₄	77		~1000	0		<1.5	<0.1 μs (> 1 kV/cm)
LAr	84	1.51	475	−0.21	0.2	0.25	<0.1 μs (> 0.3 kV/cm)
SAr	83		1000	+0.3 (6 K)		0.1	<0.1 μs (> 100 V/cm)
LKr	116	1.66	1800	−0.4	0.08	1.6	<0.1 μs (> 1.6 kV/cm)
SKr	116		3700	−0.25 (20 K)		0.98	<0.1 μs (> 1 kV/cm)
LXe	161	1.93	2200	−0.61	0.05	1.75	<0.1 μs (> 1.8 kV/cm)
SXe	161		4500	−0.46 (40 K)		1.25	<0.1 μs (> 1.3 kV/cm)

Notes: (*n-H*) normal hexane, (*iso-O*) iso-octane (trimethylpentane), (*TMP*) thetramethylpentane.

inter-phase surface may be described in terms of a one-dimensional potential as follows

$$V_1(z) = V_0 - eF_1z + eA_1, z < 0$$
$$V_2(z) = -eF_2z + eA_2, z > 0$$

(2.28)

In terms of this single dimensional representation, an electron approaching the interface can penetrate the interface if its momentum projection along z direction p_z is higher than $p_0 \approx (2m_e |V_0|)^{1/2}$ as shown in Fig. 2.6. If $p_z < p_0$, the electron can be reflected by the potential barrier back into the condensed phase, where it becomes thermalised after several collisions. Trapped under the surface, electrons stay in the thermodynamic equilibrium with the condensed dielectric and, later, can escape as a result of the thermal electron emission process.

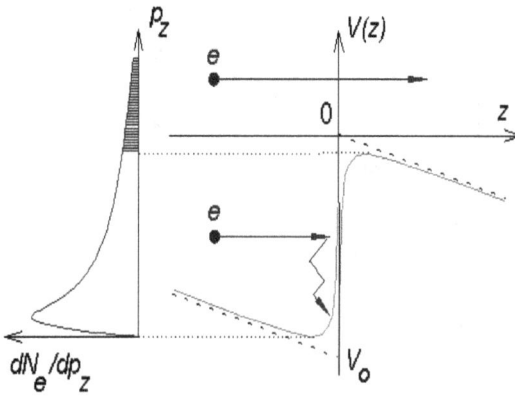

Figure 2.6 Emission of hot quasi-free electrons from condensed non-polar dielectric with negative ground state energy of electrons.

The coefficient of the quasi-free electron emission can be described as

$$K = \int\limits_{p_z > p_0} \alpha \beta f(\vec{p}) d\vec{p} \Big/ \int f(\vec{p}) d\vec{p} \qquad (2.29)$$

where α is the coefficient of the penetration of the potential barrier, β is the probability to avoid backscatter from gas, $f(\vec{p})$ is the momentum distribution function of the electrons. *Emission coefficients* measured for liquid and solid argon and xenon by Guschin *et al.* (1982) are presented in Fig. 2.7 as a function of the applied electric field strength in the condensed dielectrics.

Effective emission of quasi-free electrons has been observed from condensed noble gases, methane and some other liquid saturated hydrocarbons (Table 2.1). In general, the electron emission from non-polar dielectrics looks like a threshold process (Fig. 2.7) with a strong correlation between the observed threshold emission electric field F_0 and the energy of the ground state of excess electrons V_0 as shown in Fig. 2.8.

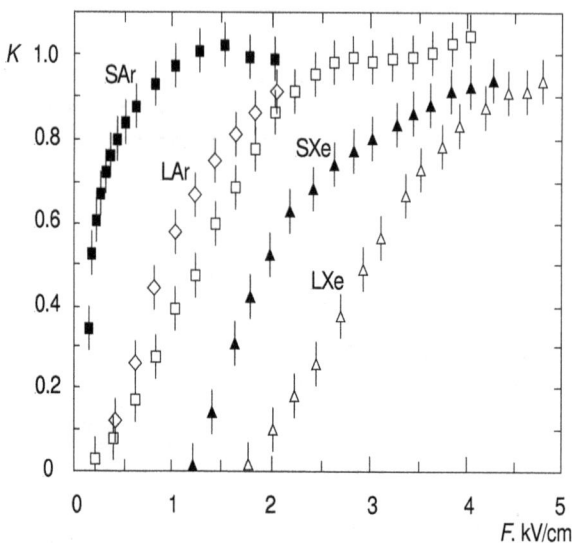

Figure 2.7 Coefficient of quasi-free electron emission from condensed argon (solid at 80 K and liquid at 90 K; squares indicate 'fast component', diamonds indicate 'slow' component) and xenon (solid at 160 K and liquid at 165 K). Redrawn from Guschin, 1982.

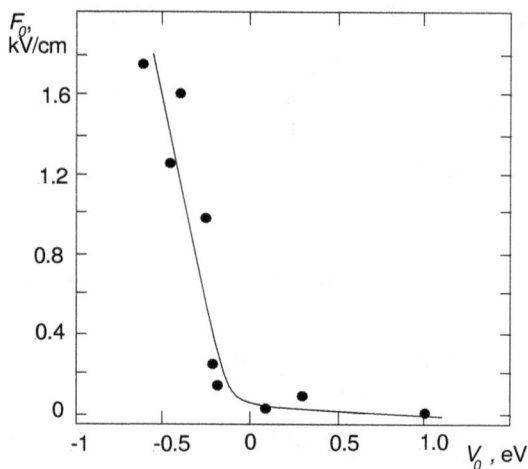

Figure 2.8 Correlation between the emission threshold and the energy of the ground state in non-polar dielectrics (Bolozdynya, 1991).

2.3.2 *Thermal electron emission from non-polar dielectrics*

In liquid saturated hydrocarbons, such as iso-octane and n-hexane and liquid argon, the value of the interface potential barrier is comparable to the kinetic energy of quasi-free electrons, when they are in thermodynamic equilibrium with the medium: $V_0 \sim k_B T$ (see Table 2.1). In this case, electrons from the high-momentum tail of the momentum distribution ($p_z > p_o$, Fig. 2.6) have sufficient energy to escape from the liquid.

Assuming the Maxwellian distribution function $f(\vec{p})$ for thermal quasi-free electrons, the probability of existence in the quasi-free state (2.26) and using formalism similar to that described in Section 2.4.1, Balakin, Boriev and Yakovlev (1977) derived the value $V_0 = -0.19 \pm 0.02$ eV and its temperature derivative $\partial V_0 / \partial T = -0.8$ meV/deg for liquid iso-octane from their emission time measurements (Fig. 2.9) and found that these values are in good agreement with the results obtained from metal-to-liquid electron photo-emission data. Thus, the thermal nature of the quasi-free electron emission from saturated hydrocarbons has been proven.

It was found (Balakin, Boriev and Yakovlev, 1977; Bolozdynya, Lebedenko *et al.*, 1978; Borghesani *et al.*, 1990) that the thermal electron emission time t_e from non-polar dielectrics with shallow electron ground state, such as liquid iso-octane and argon, is ranging between 10^{-3} sec and 10^{-6} sec and is inversely depending on the strength of the extraction electric field F, as shown in Fig. 2.10.

For noble liquids, the thermal electron emission time may be calculated in terms of Cohen-Lekner formalism as

$$t_e \sim (\Lambda / v_d) \exp\{[V_0 - 2eA_1^{1/2}(1 + A_2^{1/2} / A_1^{1/2})F^{1/2}] / k_B T\} \qquad (2.30)$$

where Λ and v_d are the electron momentum free path and the drift velocity, respectively (Borghesani *et al.*, 1990). Using this approach, Borghesani, Carugno and Santini (1991) have determined the energy of the conduction band of quasi-free electrons in liquid argon from their measurements of the electron emission time. The value found for V_0 agreed with the results obtained from metal-to-liquid electron photo-emission data.

Due to the low energy of escaping thermal electrons, backscattering of electrons in the gas phase did not play significant role in the thermal electron emission experiments.

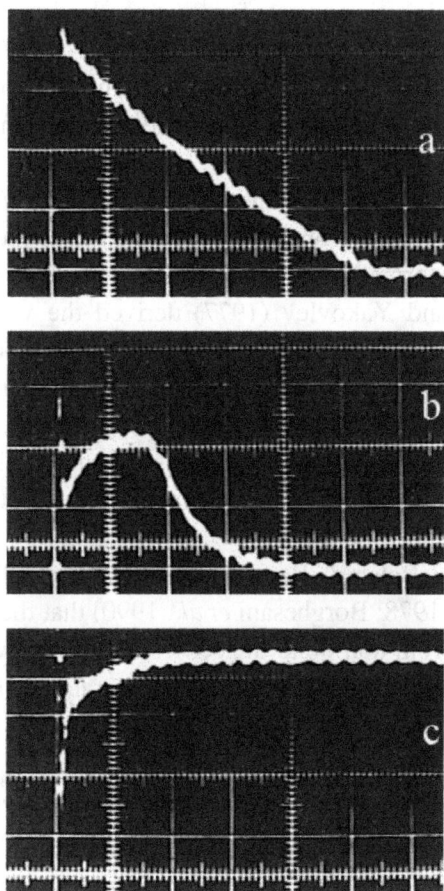

Figure 2.9 Pictures taken from the oscilloscope screen of waveforms of the electric current observed in the diode ionization chamber filled with liquid isooctane, when both electrodes (a) or only the cathode (b) or only the anode (c) are immersed in the liquid. The diameter of the electrodes is 1.5 cm, distance between electrodes is 7 mm, thickness of the liquid is 2 mm (b, c), applied voltage is 2000 V, temperature is 296 K, 0.1 μA/div vertical scale, 20 μs/div horizontal scale (Balakin *et al.*, 1977). Courtesy of B.S. Yakovlev.

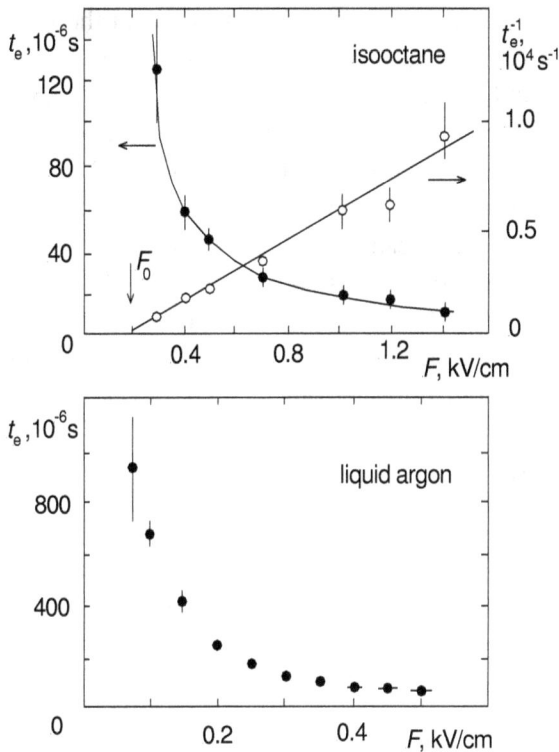

Figure 2.10 Thermal electron emission time in dependence of electrical field strength in liquid iso-octane at room temperature (top) and liquid argon at 87.4 K (bottom). Redrawn from Bolozdynya, 1986 (top) and Borghesani *et al.*, 1990 (bottom).

2.3.3 Hot electron emission from non-polar dielectrics

In heavy noble (Ar, Kr, Xe) liquids and solids with high atom polarisibility and strong interaction between atoms, electrons predominantly exist in quasi-free state, which is characterised with high drift velocities even exceeding drift velocities in the equilibrium gas phase at the same *reduced electric field F/n* where n is the density of the gas measured in $[cm^{-3}]$ units.

In condensed Kr and Xe, the potential barrier is so high ($|V_0| \gg k_B T$) that at achievable levels of purification the electron lifetime is normally much shorter than the emission time and the thermo-activated electron emission from these liquids and solids has not been observed. On the other hand, in these media it is easy to achieve the electric field strength F_c in which drifting electrons have average kinetic energy exceeding the temperature of the media and even generating electroluminescence of the liquids. As shown in Fig. 2.6, hot electrons with momentum $p_z > p_0$ can escape the condensed phase without delay. Effective and fast electron emission from these media has been observed at $F > F_c$ (Fig. 3.21). Note that electrons that are not emitted cannot continue their drift and, thus, cannot be heated by the applied electric field and because of that they become quickly (~ 1ps) thermalised.

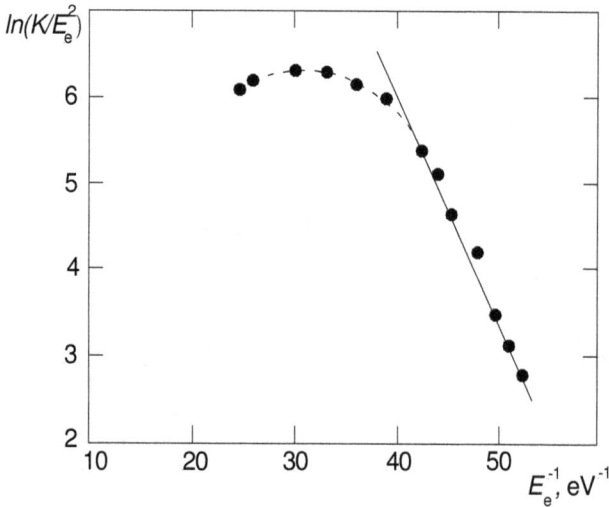

Figure 2.11 Probability K of quasi-free electron emission from liquid krypton versus the average electron energy in RLD coordinates (Bolozdynya, 1991).

Quasi-free electron emission from liquid krypton has been analysed (Bolozdynya, 1991) in terms of the basic law describing the electron emission current from hot cathodes (Section 2.1.2). Since quasi-free electrons in condensed krypton can be heated by the electric field, the Richardson–Laue–Dushman law (Eq. 2.12) can be applied and the emission coefficient can be expressed as follows

$$K = I_e / I_i = \langle E_e \rangle^2 \exp\left[-\varphi / (2/3)\langle E_e \rangle\right] \qquad (2.31)$$

where I_e and I_i are the electron emission current and electron ionization current, respectively, and $\langle E_e \rangle$ is the average electron energy in dependence on the applied electric field F. Using the measured dependences of $I_e(F)$ and $I_i(F)$ and the calculated $\langle E_e \rangle(F)$ for the electric field strength in liquid krypton, the graph of $\ln(K/\langle E_e \rangle^2)$ versus $\langle E_e \rangle^{-1}$ has been built as shown in Fig. 2.11. The slope of the linear part of this curve allows defining the work function of krypton to be $\varphi = 0.39 \pm 0.07$ eV that nicely matches the value of $V_0 = -0.4$ eV measured in other experiments (Table 2.1).

The linearity of the RLD-plot serves as evidence of the hot electron emission from heavy noble liquids. The non-linear left wing of the curve is apparent in this type of experiment due to limited electron density in the conductance band (see, for example, Chung and Yoon, 2003). In our case, this is a result of high-field saturation of the ionization yield from tracks of X-ray photo-electrons used in this study.

In general, emission of quasi-free electrons from heavy noble liquids and solids is a combination of thermal and hot electron emission. That explains the observation of two components (*fast* and *slow*) of the electron emission from liquid argon (Guschin *et al.*, 1979). The 'fast' component is associated with prompt emission of electrons heated by electric field. The 'slow' component is associated electrons that did not emit immediately, cooled down to the equilibrium thermal distribution and eventually escaped with a delay as thermal electrons.

2.3.4 *Electron emission from localised states in light noble liquids*

In dielectrics with relatively small atoms, $V_O > 0$ (liquid helium, hydrogen, and neon) and excess electrons exist in low-mobile auto-localised states – bubbles. The potential well of localised electrons should be superimposed on the potential $V(z)$, Eq. 2.18, as shown in Fig. 2.12.

The emission of electrons from bubbles in liquid helium and neon into the equilibrium gas phase has been observed in several experiments (Bruschi *et al.*, 1966, 1975; Surko and Reif, 1968; Schoepe and Rayfield,

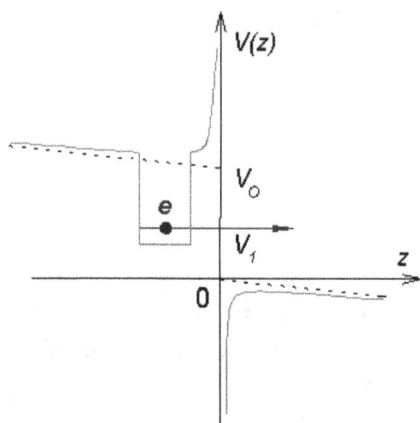

Figure 2.12 Emission of electrons from the localised state in non-polar dielectric with the positive electron ground state.

1973). The mechanism of electron emission from the localised states is a combination of thermal emission and quantum tunnelling. Superposition of the electrical potential, suppressing the electron to the free surface in the condensed medium, and the image potential, attracting the electron inside the dense dielectric, provides a shallow minimum located several tens of nanometres under the surface, depending on the value and direction of $\vec{F}$. When the electron, due to its thermal motion comes sufficiently close to the surface (less than 23 nm in liquid helium according to Anciolotto and Toigo, 1994), the potential barrier between the bubble and the surface is essentially deviated and the electron can tunnel from the localised state into the gas phase.

The process is thermo-activated: in liquid helium below the λ point, the measured current of emitted electrons decreases rapidly with decreasing temperature as $\exp(-\Phi/k_B T)$ where the effective barrier Φ/k_B ranging from 30 K to 40 K depending on the electric field. For liquid ^{4}He and ^{3}He at 1–2 K, the characteristic time of the barrier penetration (emission time) is 10–100 s at 100–10V/cm extracting electric field in the liquid (Schoepe and Rayfield, 1973). At lower temperatures, the probability of electron tunnelling is very low but at fields of > 100 V/cm the electronic bubble drifting in the superfluid helium forms a vortex, which can drag electrons through the surface without delay or overcoming any noticeable barrier (Surko and Reif, 1968, 1968b). This kind of emission may be generated by mechanical rotation of cryostat filled with super fluid helium and such way generating quantised vortex lines perpendicular to the free liquid surface (see schematics of the experiment in Fig. 4.2). Such kind of emission is temperature independent (Schoepe and Dransfeld, 1969). Note, that ionizing particles with high stopping power (alpha-particles) in superfluid helium may generate quasi-particles (rotons) with energy high enough (> 7.16 K) to produce quantum evaporation of neutral atoms (Adams *et al.*, 1998).

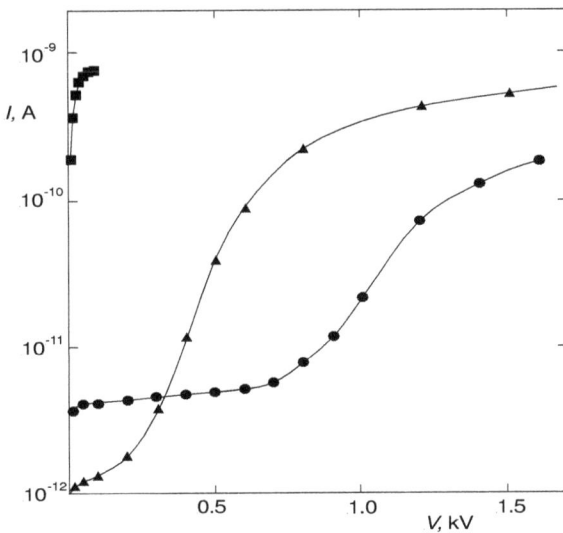

Figure 2.13 *I-V* plot of the electron emission from a 3.7 mm thick layer of liquid neon (triangles), ionization current of totally liquid-filled (squares) and vapour-filled (rounds) ionization chamber with 5 mm between the electrode gap and Am^{241} radioactive source mounted in the centre of the cathode and used as the bottom electrode at T = 35.7 K. Redrawn from Bruschi *et al.*, 1975.

Since the V_O value is very high in liquid helium, the thermal electron emission practically has no observable electric field threshold (Fig. 2.18). In contrast, the thermo-activated electron emission from liquid neon looks like a threshold process with emission current rapidly increasing above $F_0 \sim 1$ kV/cm at 35.7 K (Fig. 2.13). When tunnelling from localised states in liquid neon, electrons demonstrate the emission time in the range of 1–100 s at 1.0–0.1 kV/cm extraction electric field in the liquid (Fig. 2.14).

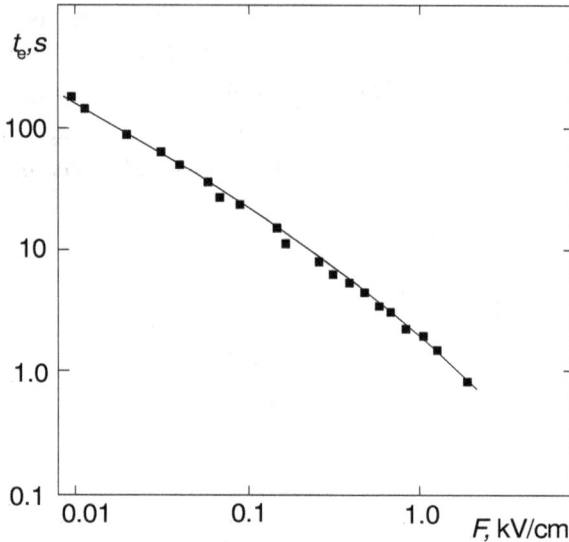

Figure 2.14 The electron emission time from liquid neon plotted against the electric field strength in the liquid at T = 35.7 K for thicknesses of the liquid layer ranged between 0.5 and 5 mm. Redrawn from Bruschi *et al.*, 1975.

2.4 Charge carrier emission due to liquid surface instability

The stability of the horizontal liquid interface under the influence of a vertical electric field attracted attention of researchers beginning from the first experiments on photo-emission from liquid dielectrics performed by Stoletow at the end of the 19th century.

2.4.1 Stability of charged liquid interface

Landau and Lifshits (1953) considered the problem formulated by Ya. Frenkel in 1935 and defined conditions for instability of the charged liquid surface due to *gravitational micro-waves*. The idea was that the charged surface experiences additional negative pressure due to applied electric field that at certain conditions may compensate the surface tension. Landau and Lifshits derived the dispersion relation for waves at the charged liquid surface as

$$\omega^2 = gk - 4\pi\sigma^2 k^2 / \rho + \tau k^3 / \rho \qquad (2.32)$$

where g is the local gravitational acceleration, τ is the surface-tension coefficient, ρ is the liquid density, σ is the surface charge density. For stability, the frequency ω must be real for all values of the wave number k, i.e. $\omega^2 > 0$, that means

$$\tau k^2 - 4\pi\sigma^2 k + \rho g > 0 \qquad (2.33)$$

This condition for the above quadratic equation is realised for the negative determinant

$$(4\pi\sigma^2)^2 - 4\rho g \tau < 0 \qquad (2.34)$$

From this equation one can find that the surface is stable if the surface density of charges is not exceeding the critical value:

$$\sigma < \sigma_0 = (\rho g \tau / 4\pi^2)^{1/4} \qquad (2.35)$$

The described model does not involve a detailed knowledge of the form of the unstable disturbances but still predicts the charge density threshold of instability with good accuracy.

Due to the image potential, electrons can form a two-dimensional layer on the surface of liquids having a positive ground state of quasi-free electrons V_0 (liquid helium or liquid hydrogen, for example). There is a certain limitation on the electron surface charge density $\sim 2 \cdot 10^9$ e/cm^2 (Brown and Grimes, 1972); the electron density at the surface of a film of superfluid helium on a metallic base may be an order of magnitude larger (Volodin, Khaikin and Edel'man, 1976). The over-charged liquid helium surface loses its stability and forms bubbles with a 0.05–0.3 mm diameter (depending on the number of electrons localised inside the bubble) diving into the liquid (Khaikin and Volodin, 1978). On reaching the anode, the bubbles lose their charge and collapse. The collapse time depends on the size of bubble and varies from $< 10^{-4}$ s for the smallest bubbles to $\sim 10^3$ sec for large ones containing 10^7–10^8 electrons. Similar effect has been observed in liquid Ne, and liquid H_2 (Troyanovky, Volodin and Khaikin, 1979).

2.4.2 Stability of non-charged dielectric liquid interface

Since emission detectors have to deal with relatively low charge carrier densities produced by radiation, it is important to consider conditions for stability of the non-charged liquid dielectric interface.

Stability of the dielectric liquid-gas interface can be broken by polarisation charges induced by the applied electric field. A surface density of polarisation charge is defined by the difference between electric fields in gas and in liquid

$$4\pi\sigma = F_2 - F_1 \qquad (2.36)$$

Then, using the critical charge density for *stability of the liquid surface* (Eq. 2.35), the critical electric field for extraction ions from the liquid with dielectric constant ε_1 and density ρ_1 into the phase with dielectric constant ε_2 and density ρ_2 can be calculated as

$$F_{20} = (8\pi)^{1/2} \frac{\varepsilon_1}{\varepsilon_1 - \varepsilon_2} \left([\rho_1 - \rho_2]g\tau\right)^{1/4} \qquad (2.37)$$

Taylor and McEwan (1965) of Kavendish Laboratory considered the stability of the liquid interface with an apparatus shown in Fig. 2.15. They have observed that the field lifts up the part of the interface below the electrode until stability of the interface is broken at the certain critical electric field F_0. Experiments have been carried out with several pairs of conductive-nonconductive liquids, water-air and mercury-air interface. They have shown that horizontal interfaces between conducting and non-conducting fluids become unstable under the action of a sufficiently great electric field; however, the non-conductor interface placed between two electrodes most often suffers electrical breakdown at a

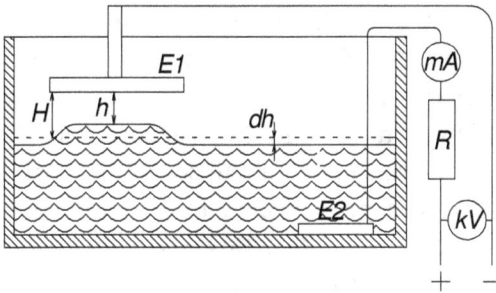

Figure 2.15 Apparatus for measurements of interphase instabilities in dielectric liquid at room temperature. Redrawn from Taylor and McEwan, 1965.

voltage gradient lower than that which can cause instability of the interface. The breakdown was probably triggered by ionization from cosmic rays or natural radiation. The exception is liquids with high dielectric constant such as water demonstrating lower critical electric fields (F_{20} = 24.6 kV/cm in air above the liquid) to be in good agreement expressed in Eq. 2.37. For comparison, the predicted critical electric field in vapour above liquid nitrogen is 47 kV/cm, above liquid xenon — 49 kV/cm, above liquid helium — 26-73 kV/cm depending on temperature in the range between 5.1 and 3.2 K. Such high strength electric fields are difficult to apply to the gas phase before breakdown.

The instability of liquid interfaces realises as *geysers* (jets) of liquid moving to opposite electrode and disintegrating into droplets (Volodin and Khaikin, 1979). At certain conditions, the liquid dielectrics may form a liquid layer at the opposite electrode. The formation of a liquid layer may continue until the equilibrium between the Coulomb force, the gravity and the surface tension is established, and then, the liquid trickles down from the top electrode towards a lower one (Mardarskii *et al.*, 2000).

2.4.3 Ion emission from dielectric liquids

Emission detectors have to deal with low density of electrons occurring due to ionization of condensed dielectrics with ionizing radiation. In this case, interaction of single charge carriers with the

interface is the most probable process. In previous sections we have
considered conditions for single electron emission. However in many
dielectric liquids charge carriers exist only as ions. Here we will consider
conditions that allow extraction of ions from condensed dielectrics.

It was shown experimentally that under influence of high extraction
electric fields, ions can be extracted from liquid nitrogen and neon
(Bruschi, Mazzi and Torzo, 1975; Boyle and Dahm, 1976) at electric
fields a few times less than that expected from the liquid surface
instability as described in the previous section but significantly higher
than that required for extraction of electrons.

Figure 2.16 presents typical data on ion extraction from liquid
helium at 4.2 K. As seen, the negative current penetrates the surface at all

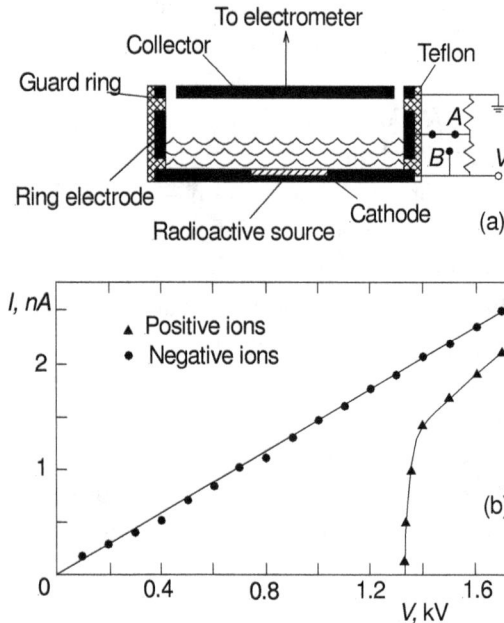

Figure 2.16 Cell for measurements of ion emission from liquid
helium at T = 4.2 K (a) and *I-V* plot for operation in mode *A* for
positive (triangles) and negative (circles) ions at 4 mm interface-
collector separation, 8 mm collector-cathode separation (b).
Redrawn from Boyle and Dahm, 1976.

voltages above 100 V. No positive current is collected until at some threshold voltage the current rises abruptly and then asymptotically approaches the value of the negative current. This threshold voltage is temperature dependent and approaches zero at the critical point of helium. In the mode B (the switch is turned down in Fig. 2.16a), the increased density of the surface charge lifts up the liquid surface in the centre of the ionization chamber similar to that observed by Taylor and McEwan (1965) as described in the previous section and shown in Fig. 2.15.

Rodionov (1969), and independently Boyle and Dahm (1976), proposed a simplified model of *droplet emission*. They assumed that micro-waves at the liquid surface cause hemispherical shaped protrusions, which under the influences of an extracting charge electric field, acting against the gravitational force and the surface tension may be separated from the surface inside a droplet. With this model, the extraction of ions from dielectric liquids may be described in the following sequence of events: a drifting ion approaches the liquid surface; the ion is captured inside the microwave; the extracting electrical force deforms the surface; a droplet containing the ion separates from the surface and begins to evaporate due to large curvature of its surface; the extracted ion departs from the interface. Using this model, Boyle and Dahm derived the formula for the critical electric field in the gas:

$$F_{BD20} = (8\pi)^{1/2} \frac{\varepsilon_1}{\varepsilon_1 - \varepsilon_2} \left([\rho_1 - \rho_2] g \tau / 3 \right)^{1/4} \qquad (2.38)$$

Calculations in the frame of this model gave results close to the experimental data on extraction of positive ions from liquid helium and ions of both signs from liquid nitrogen (Table 2.2). The critical field is temperature dependent and approaches zero at the critical point.

More accurate consideration of the liquid surface instability, then, should take in account a superposition of the applied electric field and the local electric field of charge carriers. A single charge carrier placed in the liquid dielectric and suppressed by the applied electric field to the liquid surface acts on this surface by its local electric field

$$F_{el} = \frac{e}{\varepsilon_1 z_0^2}$$ (2.39)

where z_0 is the depth of the charge-carrier localisation under the surface due to the image potential given by Eq. 2.15. This local electric field may be significantly stronger than the applied electric field F_1 and may affect the local stability of the liquid surface. If the macroscopic consideration in this case is still valid, combining Eqs. 2.15, 2.37 and 2.39, one can derive a new formula for the critical extraction electric field in the gas phase

$$F_{e20} = \left(\frac{\pi}{2}\right)^{1/2} \frac{\varepsilon_1}{\varepsilon_1 + \varepsilon_2} (\Delta \rho g \tau)^{1/4}$$ (2.40)

Equation 2.40 precisely describes the experimental data on positive and negative ion emission from liquid nitrogen and negative ion emission from liquid neon, however, gives too low values of the critical field for extraction of positive ions from liquid helium and liquid xenon. This may be explained by a change of density of the liquid in the vicinity of the ions.

Table 2.2 Critical electric field of ion extraction from some dielectric liquids.

	T, K	τ, erg/cm^2	ρ, g/cm^3	ε	ions	F_{20}^{exp}, kV/cm
N$_2$	77	8.3	0.808	1.4	+/−	19 [a]
He	3.2–5.1	0.007–0.2	0.14–0.1	1.057	+	1–4 [a]
Ne	25	5.54	1.2	1.19	−	1.8 [b]
Xe	161.36	19.3	3.1	1.93 (164 K)	+	> 16 [c]

Notes: [a] Boyle and Dahm (1976); [b] Bruschi *et al.* (1975); [c] Walters (2003).

Chapter 3

Generation of Signals in Massive Emission Detectors

Most important practical applications of emission detectors are associated with massive working media made of non-polar dielectrics. A consequence of elementary processes leading to generation of measurable signals acquired from emission detectors is the following:

1) interaction of radiation (particles) with condensed working medium and excitation generation (photons, phonons) and ionization (charge carriers) in the working medium;

2) collection and detection of primary excitations;

3) transportation of charge carriers through the bulk working media;

4) extraction of the charge carriers from the condensed medium into another medium for amplification of the ionization signal;

5) amplification and measurement of the ionization signal.

In Chapter 2, we showed that, practically speaking, only condensed noble gases and saturated hydrocarbons can be used for the construction of massive emission detectors. For this reason, we focus our consideration mostly to processes occurring in the non-polar dielectrics. Signals acquired from noble gas detectors are generated as a result of the motion of charge carriers and/or the collection of photons produced during the dissipation of absorbed energy from measured radiation as described in Chapter 2. In order to amplify signals, electron avalanche multiplication and excitation of the media by drifting electrons are often used. Both original and secondary electrons and photons can be used to acquire data from absorbed energy, transferred momentum, position and time of interactions, and for the identification of ionizing particles. In this chapter, we consider elementary processes associated with collection of charge carriers in noble gases of different aggregate states. Particular principles of operation of different detectors will be considered elsewhere.

3.1 Dissipation of energy deposited by radiation

Operations of detectors are based on interactions of detected radiation with the detector media. The interactions lead to energy transfer from detected radiation or particles to the detector medium. The art of detection technology consists of the effective transformation of the deposited energy into electronic signals that can be measured with modern electronics and connected to data-processing systems.

Charged particles and photons can directly interact with media due to the excitation of electrons in atoms. Neutral particles (neutrons, neutrinos) deposit energy via elastic and inelastic scattering on nuclei (neutrons) or electrons (neutrinos) or are absorbed due to nuclear reactions and the production of secondary particles. The energy absorbed in detector medium E_0 is originally distributed between atom nuclei E_n and electrons E_e

$$E_0 = E_n + E_e \qquad (3.1)$$

In any case, at the final stage of the dissipation of the deposited energy ionization production and detector media excitation occurs. Ionization is associated with the generation of electrons, holes or ions. Excitations can be observed as emission of photons and quasi-particles, such as phonons (in solids) or rotons (in super fluid helium).

3.1.1 Charged particles and gamma radiation

In case of the detection of charged particles or gamma radiation, the majority of the absorbed energy is deposited via electron channel: $E_0 \approx E_e$. The energy balance of the absorbed energy E_0 is given by the *Platzman equation*:

$$E_0 = N_i E_i + N_{ex} E_{ex} + N_i E_{se} \qquad (3.2)$$

where N_i is the number of electron-ion (hole) pairs produced at the average energy of *ionization* E_i; N_{ex} is the number of excited atoms at the average energy of *excitation* E_{ex}; and E_{se} is the average kinetic energy of the *sub-excitation electrons*, which is not high enough to produce ionization or excitation of atoms and eventually goes into heat. The parameters for pure noble gases are presented in Table 3.1.

Table 3.1 Parameters of the Platzman equation for heavy noble liquids in vicinity of triple point (Takahashi *et al.*, 1975).

	E_i, eV	N_{ex}/N_i	E_{ex}, eV	E_{se}, eV
Ar	15.4	0.21	12.7	5.15
Kr	13	0.08	10.5	5.50
Xe	10.5	0.06	8.4	4.45

Since the number of ionization electrons is a primary interest for detection purposes, detector media are often characterised by their ability to generate electrons

$$W_i = E_0 / N_i = E_i + E_{ex}\left(N_{ex} / N_i\right) + E_{se} \qquad (3.3)$$

For the detector media of interest, the W_i value is close to the ionization potential I of the atom. In condensed noble gases, the ratio of W_i / I is about 1.3–1.8; in molecular gases the value lies between 2.1 and 2.6, i.e. about the half of the absorbed energy is converted into ionization. In the liquid saturated hydrocarbons, due to effective electron thermalisation and electron-ion recombination, the value W_i / I is about 0.1 and no emission of photons can be observed since the most of the absorbed energy goes to heat.

In predominant atomic media, such as pure noble gases, there are no non-radiation de-excitation transitions and an essential part of the dissipated energy is transferred into high-energy UV photon emission (for example, see monograph Aprile *et al.*, 2006). In absence of the applied electric field, the recombination process enhances excitations. These processes in pure noble gases, liquids, and solids lead to the intensive prompt light emission — *scintillation* — that plays an important role in detection technique. In suggesting that one ionization act produces one photon and one excitation produces one photon, the number of scintillation photons could be described as

$$N_{sc} = N_i + N_{ex}E_{ex} = N_i\left(1 + N_{ex} / N_i\right) \qquad (3.4)$$

In practice, the efficiency of the scintillation production is characterised with the average energy required for production of the scintillation photon W_{sc} which is connected to W_i by the following:

$$W_{sc} = E_0 / N_{sc} = W_i\left(1 + N_{ex} / N_i\right) \qquad (3.5)$$

The measured values W_i, W_{sc} and ionization potentials I for pure condensed noble gases are presented in Table 3.2.

Table 3.2 Properties of liquid noble gas detector media near triple point (extracted from Table 2.4 of Aprile, Bolotnikov, Bolozdynya, Doke (2006), see references therein).

Liquid	I, eV	$h\nu_m$, eV	W_i, eV	W_{sc}, eV	F
He	24	15.5; 20	41 (gas)	24	
Ne	25	16.02	29 (gas)	25	
Ar	13.4	9.57	23.6±0.3	19.5±1	0.107
Kr	11.55	8.42	20.5±1.5	15	0.057
Xe	11.67	7.02	15.6±0.3	13.8±0.9	0.041

High-energy electromagnetic particles (with energy in order of or above 1 GeV) passing through high-Z dense media generate electromagnetic showers. It would be too much to consider such applications for emission detectors in this publication. Interested readers are directed to the fundamental review by Fabian and Gianotti, 2003 for further explanation.

In condensed noble gases, the value W_i is measured in ionization chambers from the analysis of the saturation of the *ionization curve* or extrapolation of the *ionization yield* in the high-electric field region (see, for example, Fig. 3.1 and an excellent review by Doke, 2005).

In molecular liquids no saturated currents have been observed at applied electric fields up to 100 kV/cm (Fig. 3.2). However, a significant *free ion yield* can be observed in the absence of any electric field. Historically, the free ion yield G_{fi} is defined as the number of electron-ion pairs per 100 eV of the absorbed energy. In general, $G_{fi}(F)$ depends

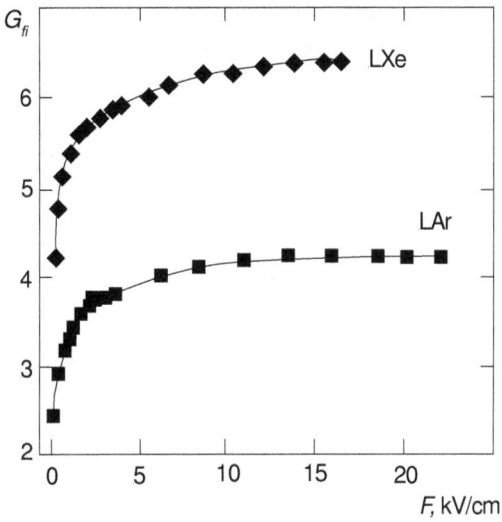

Figure 3.1 Ionization curves in liquid noble gases near triple point irradiated with relativistic electrons. Redrawn from Doke, 1981.

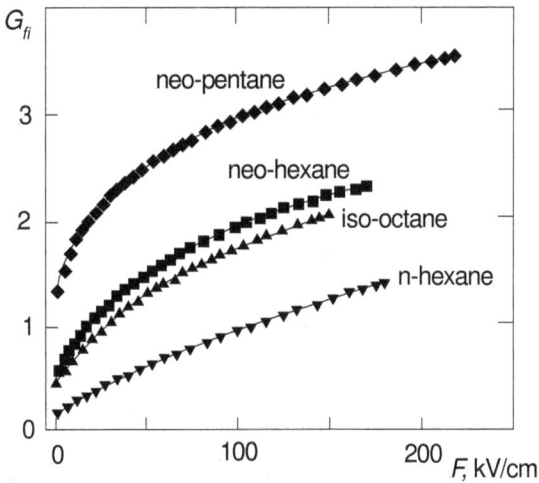

Figure 3.2 Ionization curves in liquid saturated hydrocarbons at room temperature irradiated with gamma radiation. Redrawn from Schmidt, 1997.

on media, temperature and type of radiation. For example, in liquid n-hexane irradiated with gamma-radiation at room temperature $G_{fi}(F = 0) = 0.14$; in neo-pentane at the same conditions $G_{fi}(F = 0) \approx 1.3$. The free ion yield of liquid n-hexane decreases from 0.14 at 1 bar pressure to 0.10 at 2 kbar pressure. In the limit of the extremely high electric fields, the ion yield is defined as the *total ionization yield*

$$G_{tot} = \lim_{E \to \infty} G_{fi}(F) \qquad (3.5)$$

The connection between the total ionization yield G_{tot} and the energy required for electron-ion production is

$$G_{tot} = 100eV / W_i \qquad (3.6)$$

Details about the processes of ionization and recombination of electrons and ions in *saturated hydrocarbons* may be found in Schmidt (1997).

3.1.2 Nuclear recoils

In the process of detection of neutral particles, scatter from nuclei becomes a dominant channel of the deposited energy dissipation, and $E_0 \approx E_n$. For slow moving particles, the probability of scattering may be increased if the de Broglie wavelength of the particle is comparable to the size of nuclei. This effect of coherent scattering has been considered for detection of WIMPs and low-energy neutrino in noble liquid detectors (see Chapter 7).

Detection of low energy nuclear recoils has two specifics. First, the cross-section of the scatter on large atoms such as Xe is affected by interference of waves scattered from different nucleons (nuclear form-factor). Second, the effective recombination along tracks of recoils leads to greatly reduced ionization yield (by a factor of 5–10) or an abnormally low W_i value, relative to electrons at the same deposited energy (quenching effect).

Nuclear form factor

In case of the elastic scattering of a neutral particle on a nucleus of mass M_N with energy transfer E_R, following simple kinematics consideration, the momentum transfer can be defined as

$$q = \left(2M_N E_R\right)^{1/2} \tag{3.7}$$

At very low energy transfers, the wavelength of the scattering particle λ may no longer be large compared to the nuclear radius r_N

$$\lambda = h / q \sim r_N \tag{3.8}$$

where h is Plank's constant. In that case, the interference of scattered waves from different nucleons in the nucleus affects the total cross-section of the scattering process. In formal language, this effect can be reflected as a correction of the experimentally measured scattering cross-section with *form factor* $F(q)$

$$\sigma(q r_N) = \sigma_0 F^2(q r_N) \tag{3.9}$$

where σ_0 is the cross-section at zero momentum transfer. In practice, the form factor results in a loss of detector sensitivity for certain energy depositions.

There is significant literature devoted to calculations of form factors for different detector media and scattering particles (see Engel, 1991; Lewin and Smith, 1996; Ressel and Dean, 1997). For the following consideration, it is important to have in a mind the form factors for atoms that can be used in massive emission detectors (Fig. 3.3).

Quenching effect

In the both scintillation and ionization detectors there is a significant difference between numbers of generated photons or charge-carriers between electrons and nuclear recoils at the same deposited energy. This effect is a result of a higher ionization density and very effective recombination along the nuclear recoil tracks. The ratio of empirical values of ionization (number of electron-ion pairs) or scintillation (number of excitations) by nuclear recoil to similar value measured for electron (or gamma-ray) interactions is called the *quenching (Lindhard) factor q*. Lindhard *et al.* (1963) defined the quenching factor as the ratio of energy given to the electronic excitation to the total energy deposited by the detected particle into the detector medium.

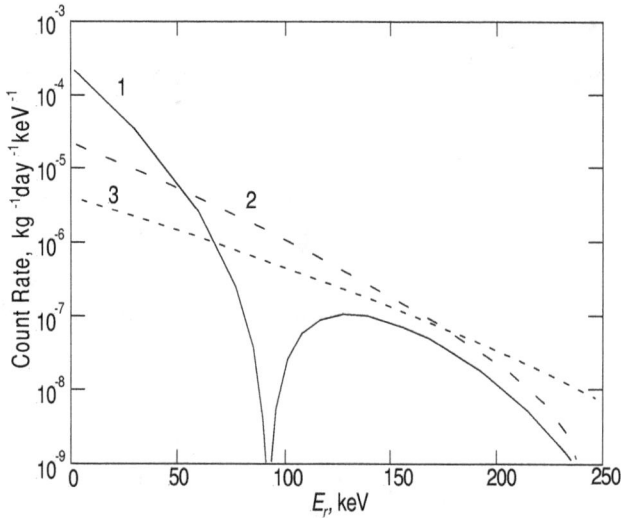

Figure 3.3 Form factors of ^{132}Xe (1), ^{40}Ar (2) and ^{20}Ne (3) as a function of the recoil energy for spin independent scattering. Redrawn from Arisaka and Smith, 2009.

Table 3.3 Nuclear quenching measured for nuclear recoils in popular detector materials.

Medium	Nucleus	q	E_R, keV	Reference
LXe	Xe	0.2	50-100	Arneodo *et al.*, 2000
		0.22±0.01	40-70	Akimov *et al.*, 2002
		0.45±0.12	10-60	Aprile *et al.*, 2005
NaI(Tl)	Na	0.3	15	Spooner *et al.*, 1994
	I	0.08	27	
CaF(Eu)	Ca	0.08		Davies *et al.*, 1994
	F	0.12		
Si(Li)	Si	0.30	3.2-21	Gerbier *et al.*, 1990
Ge	Ge	0.25	3-20	Messous *et al.*, 1995

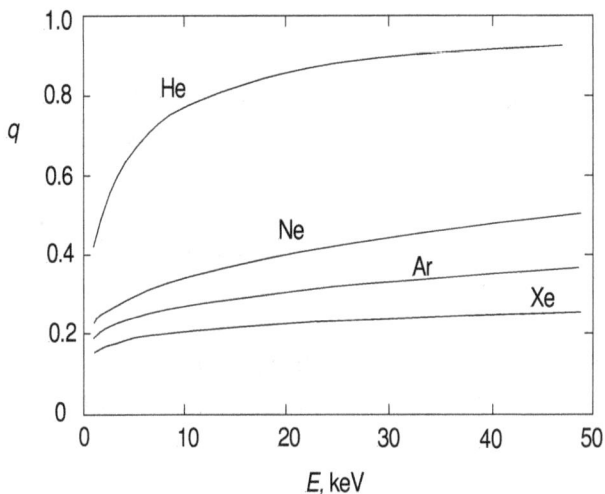

Figure 3.4 The nuclear quenching factor calculated for recoil ions as a function of recoil energy in Linhard model for He, Ne, Ar and Xe (top to bottom). Redrawn from Hitachi, 2007.

For bolometric detectors the total deposited energy is measured without a distinction between optical, ionization and heat signal generation, and so $q = 1$. The quenching factors for some popular condensed detector media are listed in Table 3.3. The quenching factors for noble gases in the range of 1–50 keV are shown in Fig. 3.4. The quenching factor decreases as the mass of the media nuclei increase.

3.1.3 Fluctuations of ionization yield

Transformation of the energy deposited by radiation into a number of charge carriers allows the measurement of this deposition by counting the number of the produced charge carriers (electron-ion pairs) N_i:

$$E_0 = W_i N_i \qquad (3.10)$$

Accuracy of the energy measurement is affected by natural fluctuations of N_i or ionization yield. In 1947, U. Fano showed that the *fluctuations of ionization* is different from Poisson distribution and may be represented by the following way

$$\delta^2 = \left\langle \left(N - N_i\right)^2 \right\rangle = F N_i \qquad (3.11)$$

where the Fano factor F depends on properties of the detector medium and on process of the interaction of the measured radiation with this medium. At $F = 1$ Eq. 3.11 represents a *Poisson distribution*; for absolutely identical ionization acts, F would be equal to 0. In ionization detectors are filled with pure noble gases and irradiated by particles with minimal ionizing power, $F < 1$ (see Table 3.2). For highly ionizing particles and low energy depositions, quenching is observed (see previous section) and so the Fano factor $F > 1$ (Alkhasov *et al.*, 1967) has shown that, at first approximation, it depends on the ratio between the numbers of excitations and ionizations

$$F \approx \frac{N_{ex}}{N_i} \cdot \left(1 + \frac{N_{ex}}{N_i}\right) \cdot \left(\frac{\langle E_{ex} \rangle}{W_i}\right)^2 \qquad (3.12)$$

More details on calculations of the Fano factor can be found in Doke's review (2005) and in Aprile *et al.* (2006).

The ability of detectors to measure deposited energy may be characterised with *energy resolution*. Normally, the energy resolution is measured as the width of the peak of the ionization signal distribution acquired from the detector absorbing monoenergetic ionizing particles. The most common way to define the energy resolution is to measure the *full width at half maximum* (FWHM) of the measured peak. In practice of measurements of deposited energy, FWHM of the energy deposition peak ΔE depends on several independent factors that can be broken down in three terms

$$\Delta E = \left(\Delta E_0^2 + \Delta E_{el}^2 + \Delta E_c^2\right)^{1/2} \qquad (3.12)$$

where ΔE_{el} is the noise of electronic readout, ΔE_c is associated with fluctuations of charge carrier collection including emission process, and the first term is the intrinsic characteristic of the detector medium depending on fluctuations of the ionization (or excitation for scintillation detectors) yield

$$\Delta E_0 = 2.355 (F E_0 / W_i)^{1/2} \qquad (3.13)$$

Table 3.4 Energy and position resolution of noble gas emission detectors.

Media	Size, cm	$\Delta E / E$ % FWHM (E, keV)	Δx, mm	Δz, mm	Reference
S/G Xe	Ø22·0.5	15 (122)	1.5		Egorov *et al.*, 1983
L/G Ar	Ø2.5·0.4	25 (5486)			Anderson *et al.*, 1987
L/G Xe	Ø(5-10-5)· ·8.4	42 (60) 28 (122) 17 (662) 12 (1836)			Yamashita *et al.*, 2004
L/G Xe	Ø2·2			0.5	Guschin *et al.*, 1882
L/G Xe	Ø5·1.5	18 (50ee)		0.4	Afanasiev *et al.*, 2003
L/G Xe	Ø20·15	44 (122)	7	< 1	Angle *et al.*, 2007

Note: (L) liquid; (G) gas; (S) solid; (L/G) emission detector with liquid-gas filling; (ee) electron equivalent.

Both the W_i value and the Fano factor are affected by fluctuations in number of charge carriers generated by detected particles. In gas detectors at energy depositions of > 50 keV, the tendency is that the Fano factor is a linear function of W_i (Bronic, 1992). However, at higher ionization density, such as happens in nuclear recoil tracks in condensed dielectrics, the Fano factor degrades (increases) much faster than the reduction of the value of W_i, leading to degradation of the energy resolution. This effect reduces the ability of liquid xenon emission detectors to distinguish electron and nuclear recoils (see Section 7.2). Measured energy resolution and position resolution of emission detectors are given in Table 3.4.

3.2 Transfer of charge carriers through non-polar dielectrics

Drifting is the motion of charge carriers under the influence of an electric field. In the absence of external forces, electrons in a gas of temperature T move with a Maxwellian energy distribution with an average value of kT (about 0.025 eV at room temperature). Under the action of an electric field $\vec{F}$, carriers acquire a net motion in the direction of the electric field with stationary drift velocity v_d averaged from instantaneous velocities: $v_d = \langle v(t) \rangle$. At weak enough electric fields, when the carriers elastically collide with atoms and molecules of the drift medium, the *drift velocity* is proportional to the electric field:

$$\vec{v}_d = \mu\vec{F} \qquad (3.14)$$

The proportionality factor μ is called the *mobility*. The mobility of electrons is a constant at low electric fields. The mobility of ions is a constant in practically all important cases. The probability of inelastic collisions of ions is valuable at instantaneous velocities comparable with the electron velocity in the atom: $v(t) \sim 10^8$ cm/s, i.e. ions, should have energies greater than 10 keV to collide inelastically. This is practically impossible at applied electric fields.

The carrier distribution in the volume of the ionization detector is mostly non-uniform. The gradient of concentration spreads carriers throughout the volume. This process is called *diffusion*. In radiation detection, the density of carriers in detector media is often sufficiently small and one can neglect Coulomb interactions between them. The current density of carriers can then be described as

$$J = -D \cdot \nabla n + nv_d \qquad (3.15)$$

At weak electric fields and small carrier concentrations, the mobility doesn't depend on the electric field $\vec{F}$ and can be expressed with *Nernst–Einstein equation*

$$\mu = eD / kT \qquad (3.16)$$

'Weak electric field' means that carriers exist in thermodynamic equilibrium with the medium and are said to be thermalised. Ions are nearly always in this state. If charge carriers are not thermalised, the mobility depends on the electric field and the drift cannot be described in the above simple model. The diffusion coefficient is then a tensor: D_T in directions perpendicular to the electric field and D_L in the direction of the drift. Note that the magnetic field introduces more anisotropy in the diffusion coefficient tensor. For more details, see McDaniel and Mason, 1973; Huxley and Crompton, 1974; Borghesani, 2007.

3.2.1 Drift of electrons

Electrons behave very differently in different condensed noble gases at different conditions. Obviously, their behaviour is classified via the value of mobility in zero limit of the electric field, as discussed above in Section 2.3.1.

Electrons with mobility $\mu_0 > 10$ cm^{2}V^{-1}s^{-1} are classified as *quasi-free*. Such high mobility can be observed in heavy noble liquids and solids and saturated hydrocarbons with spherically shaped molecules such as methane, thetramethylpentane (TMP), tetramethylsilane (TMSi), trimethylpentane (iso-octane), neopentane (NP).

Electrons with mobility $\mu_0 < 0.1$ cm^{2}V^{-1}s^{-1} are localised in deep traps or density fluctuations that develops such traps. Such low mobile electrons form vacuum bubbles in light noble liquids (LHe, LNe) and hydrogen.

The third class is a transition class where different states are possible with different probabilities. This kind of electron mobility is specific for normal saturated hydrocarbons with long chain-like molecules such as n-alkanes.

For all types of electron mobility, the drift velocity is proportional, as in Eq. 3.14, to the applied electric field at low fields (< 100 V/cm). A deviation from this proportionality occurs at elevated electric fields (Fig. 3.5). This effect is associated with *heating electrons* whose energy distribution becomes decoupled from thermodynamic equilibrium with the medium. At this range, the saturation effect in $v(F)$ dependence is observed in noble liquids and solids.

At high drift fields, the kinetic energy of electrons may exceed the excitation energy of noble atoms and electroluminescence of condensed heavy noble liquids is observed.

Quasi-free electrons

In heavy noble liquids and solids (Ar, Kr, Xe), because of high atom polarisibility and strong interaction between atoms polarised in the electron field, electrons predominantly exist in quasi-free state, which is characterised with high drift velocities, exceeding those in the gas phase at the same reduced electric field F/n (Fig. 3.5). Heating electrons using the electric field in the liquid begins for less than the values of F/n parameter than in the gas. The mobility in the gas remains constant up to $F/n = 0.02$ Td (*Townsend's unit*: 1 Td $= 10^{-17}$ V·cm^2), while the mobility in the liquid begins to decrease at $F/n = 10^{-4}$ Td. This is a result of small electron scattering cross-sections in the liquids compared to the cross-section in gas. Electron drift velocity in the liquid is limited for high electric fields and approaches to drift velocity in the gas as the field increases. The characteristic energy in the liquid is larger than the electron mean energy; the difference is smaller in the liquid. The addition of molecular impurities increases electron drift velocity at electric fields > 1 kV/cm (see below, Section 3.2.5).

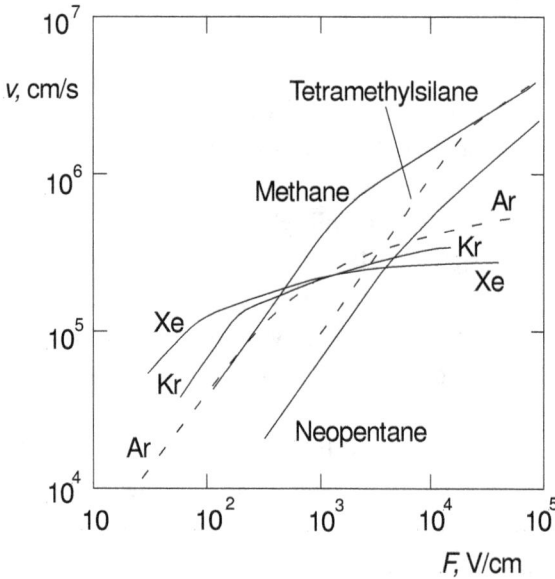

Figure 3.5 Electron drift velocity versus applied electric field in high-mobility liquids. Redrawn from Sowada, Bakale, Yoshino and Schmidt, 1975.

The theory of *hot electrons* in condensed noble gases was developed by Lekner (1967, 1968) and Cohen and Lekner (1968) and is still used today (Atrazhev *et al.*, 2005; Sakai, 2005) to explain the dependence of the drift velocity, mobilities, and diffusion coefficients on electric field in pure condensed noble gases and in the presence of molecular admixtures (Yoshino, Sowada and Schmidt, 1976). The theory is based on the formal consideration of the electron drift as a consequence of single elastic scatters with effective potentials of the muffin-tin type defined by Lekner and taking into account the interference of scattered electron waves. The *Cohen-Lekner theory* accurately describes transport properties of electrons in liquid argon near the triple point. However, the theory's predictions for other noble liquids are not as good (see Gryko and Popielawski, 1977).

Another approach was proposed by Ascarelli (1979), who suggested that the observed variation of the drift velocity with the electric field is a result of the time spent by the electron in *shallow traps* rather then the effect of energy transfers during collisions. This approach easily explains similarities in variations of the dependence of the drift velocity on the electric field when different molecular admixtures are added as well as the square-root field dependence of the enhancement of the drift velocity. A similar model of part-time free electrons occasionally trapped in density fluctuations is used to describe the field dependence of the electron drift velocity $v(F)$ in saturated liquid hydrocarbons (Minday, Schmidt and Davis, 1971). Ascarelli's approach may be useful in explaining the influence of density fluctuations on measured values of the Fano factor, temperature variations of the electron mobility in liquids, mobility of holes in noble solids and density effects in pressurised xenon. However, the similarity in the behaviour of electrons in noble liquids and solids can be better understood in Cohen–Lekner's formalism.

Localised electrons

In condensed noble gases with $V_0 > 0$, electrons usually demonstrate low mobility (see Table 2.1). Moreover, in superfluid ^{4}He, electrons were found to be even less mobile than positive ions. Low mobile or localised electrons have also been observed in normal liquid helium, liquid neon and liquid hydrogen (for references, see review of properties of localised electrons by Khrapak, Schmidt and Illenberger, 2005). Localisation of electrons was observed in cold high density ($n > 10^{21}$ cm^{-3}) helium gas as well (Levine and Sanders, 1967). It was found that due to the strong electron (small atom) exchange repulsion, a void or bubble of macroscopic size is created around an electron in condensed light noble gases. In the applied electric field, the bubble moves as a single entity with an electron inside with total mobility of about 10^{-2}–10^{-3} cm^{2}V^{-1}s^{-1}.

In normal liquids, the drifting *electron bubble* experiences hydrodynamic resistance and the mobility can be calculated according to the Stokes's law (Eq. 2.23). In normal liquid ^{4}He at the boiling point, the electron bubble radius is about 2.2 nm; in liquid neon around its triple point it is about 0.7 nm; in liquid hydrogen it is ~ 1 nm at 19 K. In superfluid helium, mobility of localised electrons is effectively increased but still remains smaller than that of positive ions.

3.2.2 Drift of ions

In liquid dielectrics, ion mobility is usually about 10^{-2}–10^{-4} cm^{2}V^{-1}s^{-1}, which is close to the mobility of ions in electrolytes. It is a thousand times smaller than the ion mobility in noble gases, and a million times smaller than the mobility of quasi-free electrons in heavy noble liquids. The *ion mobility* depends on the viscosity of the liquid as $\mu \sim \eta^{-\alpha}$, where the parameter $\alpha = 1$–2 for different ions and $\alpha \approx 1$ for atomic ions and quasi-spherical molecular ions. In the latter case ($\alpha \approx 1$), the mobility of ions follows Stokes's law (Eq. 2.24).

Mobility of positive ions in liquid helium can be described by Stoke's law in the suggestion (Atkins, 1959) that the ions drag about 50 He^4 atoms. Atkins assumed that electrostriction effects increase the liquid density around the ion. The size of this ion complex (*Atkins's ice ball*) is about 10–15 a_0 (where $a_0 = 5.3 \cdot 10^{-9}$ cm is the *Bohr radius*). Experimental data on ion and hole mobility in condensed non-polar dielectrics are presented in Table 3.5. Diffusion coefficients of ions may be calculated with the Nernst–Einstein formula (Eq. 3.16) using the tabulated values of mobility.

As seen in Table 3.5, there are significant differences (factor 2–3) in the mobility of *negative* and *positive ions* in non-polar dielectrics. The positive ion mobility in noble liquids is governed by the formation of a solid shell around the ion at lower temperatures. In the case of negative ions, increasing density around the ion is prevented by the repulsive interaction between the lone electron and the electron shells of the neutral atoms (Schmidt *et al.*, 2005).

3.2.3 Drift of holes

Positive holes were observed in noble solids (Le Comber, Loveland and Spear, 1975) and liquid xenon (Hilt, Schmidt and Khrapak, 1994). As seen in Table 3.5, holes demonstrate significantly higher mobility than positive ions in the liquids but approximately five orders lower mobility than electrons in the solids. The magnitude of the *hole mobility* and its temperature dependence can be described via the hopping model of charge carrier transport. In this model, the charge propagates as a result of thermally activated jumps from one trap to another with the mobility given as

$$\mu = \frac{eb^2\omega}{k_BT} \tag{3.17}$$

where b is the average jump distance and ω is the jumping frequency, which can be expressed as

$$\omega = P\frac{\omega_0}{2\pi}\exp\left(-\frac{E_a}{k_BT}\right) \tag{3.18}$$

where P is the tunnelling probability for the case where adjacent trapping sites have the same energy level, E_a is activation energy, and ω_0 is the phonon frequency.

The hole may be self-trapped in a potential well between two rare atoms, produced by the lattice and electronic distortions. This formation called a *polaron* is similar to the molecular ion R_2^+. The localised hole will occasionally tunnel to a neighbouring atom, where it will re-establish an R_2^+ formation. A great dataset of hole transport properties could be found in a classic review of hole transport in solids (Spear and Le Comber, 1977) and in a recent review of *transport properties* of localised electrons, ions, and holes (Khrapak, Schmidt and Illenberger, 2005).

Table 3.5 Mobility of ions and holes in condensed noble gases (see Aprile *et al.*, 2006; and Schmidt, 1997, and references therein).

	T, K	μ_+, cm^{2}V^{-1}s^{-1}	μ_-, cm^{2}V^{-1}s^{-1}
Methane	111	$2 \cdot 10^{-3}$	
Ethane	110	$1.8 \cdot 10^{-4}$	
Tetramethylsilane	296	$9 \cdot 10^{-4}$	
Neopentane	296	$4.3 \cdot 10^{-4}$	
^{3}He liquid	1.21	7.65	3.64
^{3}He liquid	2.94	9.36	3.50
^{4}He liquid	0.371	$5.19 \cdot 10^4$	540
^{4}He liquid	0.510	6420	209
^{4}He liquid	1.132	2.57	1.51
^{4}He liquid	2.20	0.0472	0.0326
^{4}He liquid	4.16	0.0470	0.0196
^{4}He liquid	5.18	0.0376	0.0157
Ne solid	25	0.0105 (holes)	
Ar solid	84	0.023 (holes)	
Ar liquid	90.1	$6.61 \cdot 10^{-4}$	$3.6\text{-}15 \cdot 10^{-4}$
Ar liquid	111.5	$12.2 \cdot 10^{-4}$ (Ar$_2^+$)	
Ar liquid	145.0	$26.1 \cdot 10^{-4}$ (Ar$_2^+$)	
Kr solid	116	0.04 (holes)	
Kr liquid	120	$6.45 \cdot 10^{-4}$	
Kr liquid	145.0	$6.67 \cdot 10^{-4}$ (Kr$_2^+$)	
Kr liquid	168.5	$12.19 \cdot 10^{-4}$ (Kr$_2^+$)	
Kr liquid	184.3	$10.6 \cdot 10^{-4}$ (Kr$_2^+$)	
Xe solid	157	0.020 (holes)	
Xe solid	161	0.018 (holes)	
Xe liquid	161	$35 \cdot 10^{-4}$ (holes)	
Xe liquid	230	$46 \cdot 10^{-4}$ (holes)	
Xe liquid	280	$41 \cdot 10^{-4}$ (holes)	
Xe liquid	162	$2 \cdot 10^{-4}$ (TMSi$^+$)	$7 \cdot 10^{-4}$ (O$_2^-$)
Xe liquid	192	$3 \cdot 10^{-4}$ (TMSi$^+$)	$10 \cdot 10^{-4}$ (O$_2^-$)
Xe liquid	162	$2.4 \cdot 10^{-4}$ (^{226}Th$^+$)	
Xe liquid	163.0	$1.33 \cdot 10^{-4}$ (^{208}Tl$^+$)	
Xe liquid	184.2	$2.85 \cdot 10^{-4}$ (Xe$_2^+$)	
Xe liquid	192.1	$3.17 \cdot 10^{-4}$ (Xe$_2^+$)	

3.2.4 Lifetime of charge carriers

The lifetime of charge carriers in non-polar dielectrics is limited by recombination, capture by electronegative impurities or structure traps (in solids), and attachment to electrodes or insulators due to chemical or polarisation forces. In an effective emission detector, the electron lifetime should exceed the electron drift time needed to collect electrons.

Lifetime of electrons in liquids

During the course of collecting with an applied electric field, drifting ionization electrons may be lost and thus no longer available for signal generation. For example, inelastic collisions between electronegative impurities and quasi-free electrons may result in attachment, in which the electron is captured by the neutral atom or molecule AB, producing a negative ion.

There are three major types of electron *attachments*:

1) *radiative attachment*

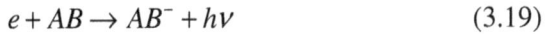

$$e + AB \rightarrow AB^- + h\nu \qquad (3.19)$$

where $h\nu$ is the photon emission;

2) *dissociative attachment* (two options)

$$e + AB \rightarrow AB^* + e \rightarrow A^+ + B^- + e \qquad (3.20)$$

$$e + AB \rightarrow AB^- \rightarrow A + B^- \qquad (3.21)$$

where AB^* is the excited state of molecule AB;

3) and three-body attachment in a two-stage *Block-Bradbury reaction*

$$e + AB \leftrightarrow AB^*$$

$$AB^* + X \rightarrow AB^- + X \qquad (3.22)$$

where X is the molecule or atom representing the majority of the medium population that plays the role of the third body carrying away the binding energy of the electron and electronegative molecule AB with positive *electron affinity*. The positive sign of the electron affinity means electronegativity of the atom or molecule, i.e. that the suitable negative ion is stable. Examples of *electronegative* atoms and molecules are tabulated in Table 3.6. Noble gases and alkali metals have negative electron affinities.

Table 3.6 Electron affinities of some atoms and molecules and thresholds of dissociative attachment to electronegative molecules (Massey, 1950; Gurvich *et al.*, 1974).

	Electron Affinity, eV	Threshold of dissociative attachment (T = 296 K), eV
F_2	2.96	~ 0
NO_2	2.43	
SO_2	1.097	
SF_6	0.6	~ 0
O_2	0.44	3.65
N_2O	0.22	0.1
NO	0.024	7.5
NH_3	< 0	5.3
H_2O	< 0	4.36
H_2	< 0	3.75
CO_2	< 0	3.99
CO	< 0	9.7
N_2	< 0	9.7

Radiative attachment (Eq. 3.19) often has a small cross-section and has no practical importance. Dissociative attachment (Eqs. 3.20 and 3.21) is a threshold process. In Eq. 3.21, electrons may be captured even by non-electronegative atoms. In condensed dielectrics, the basic mechanism of the electron attachment is a three-body reaction (Eq. 3.22) in collisions from electronegative impurities. The rate of this process may be described as

$$\frac{dn_{AB}}{dt} = -k_3 n_{AB} n_X n_e \qquad (3.23)$$

where k_3 is the reaction constant, n_{AB}, n_X and n_e are densities of electronegative impurity, atoms of the medium, and electrons, respectively. If k_3 does not depend on the densities, the two-stage reaction (Eq. 3.22) can be simplified to the one-step reaction

$$e + AB + X \rightarrow AB^- + X \qquad (3.24)$$

Then, the time-dependent concentration of drifting electrons can be described as

$$n_e(t) = n_0 \exp(-k_3 n_{AB} n_X t) = n_0 \exp(-t/t_c) \qquad (3.25)$$

where t_c is the lifetime of electrons. Measuring the electron lifetime, the *rate constant* of the three-body electron capture reaction can be defined as

$$k_3 = (n_{AB} n_X t_c)^{-1} \qquad (3.26)$$

Since the oxygen admixture limits the electron lifetime in many practical cases, Table 3.7 presents the attachment constants for thermalised electrons to oxygen in popular gases at normal conditions.

Table 3.7 Constants of three-body attachment [10^{-30} cm^{6}s^{-1}] to oxygen of thermalised electrons in gases (Shimamori and Hatano, 1977).

	He	Ne	Ar	Kr	Xe	N_2	CH_4	H_2	CO_2	H_2O
k_3	0.033	0.023	0.05	0.05	0.085	0.085	0.34	0.48	3	14

The electron attachment is often characterised as a *free drift path* L_c, associated with the electron life via the electron drift velocity as

$$L_c = v_d / t_c \qquad (3.27)$$

The rate constants for electron attachment to SF_6, N_2O and O_2 in liquid argon at 87 K and in liquid xenon at 165 K were measured as a function of drift field by Bakale, Sowada and Schmidt (1976) (Fig. 3.6). It was found that the rate constant for attachment to SF_6 is extremely high at 10V/cm, but decreases with increasing field strength by almost an order of magnitude. The rate constant for electron attachment to O_2 is a factor of 10^3 smaller and also decreases with increasing field strength in the range of 0.1–10 kV/cm.

The behaviour of the rate constant of the electron attachment to N_2O is most unusual in that the rate constant increases with increasing field strength by more than an order of magnitude. The rate constant for electron attachment to CO_2 in liquid argon also has been reported to increase at high electric drift fields (Barabash *et al.*, 1985). The similar behaviour of the attachment rate constant was observed for an unidentified residue impurity in liquid krypton (Fig. 3.7) and in liquid

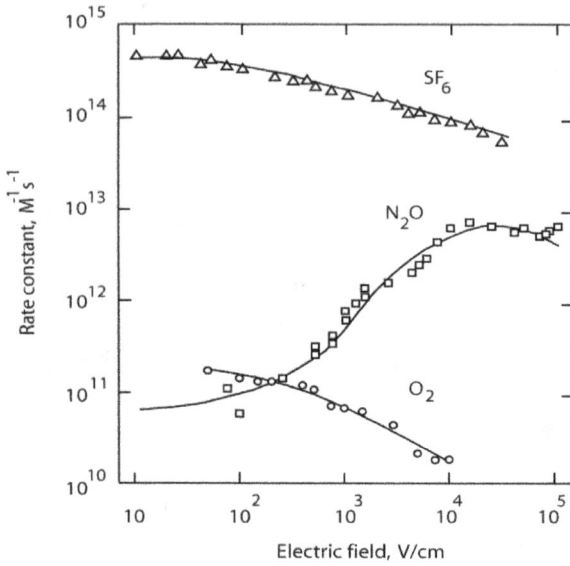

Figure 3.6 The rate constant of the electron attachment to electronegative molecules in dependence on the electric field strength. Redrawn from Bakale, Sowada and Schmidt (1976).

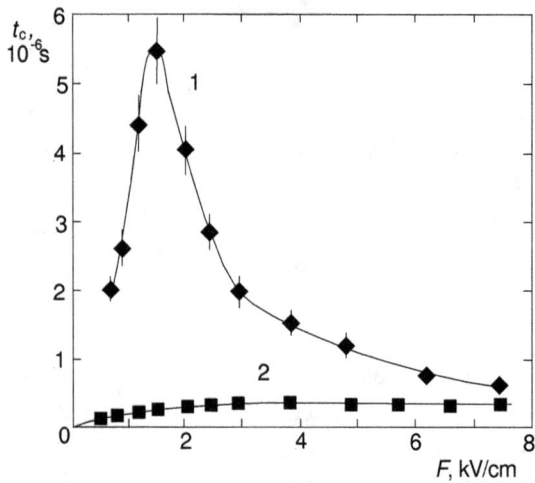

Figure 3.7 Electron lifetime in liquid krypton at 120 K in dependence on applied electric field: sample purified with hot Ca getter (1), mixed with $10^{-7}O_2$ (2). Redrawn from Bolozdynya and Stekhanov, 1984.

xenon (Obodovsky and Pokachlov, 1980; Kirilenov and Konovalov, 1981). Note that in both experiments xenon was purified in gas phase by the hot calcium getter. H. Zaklad (1971) suggested that the absorption of O_2, N_2, H_2O molecules by the calcium getter may play a role of catalyst for production of N_2O, NO, NH_3 impurities. The combination of a few types of chemical absorbers may help solve this problem. It is also shown that spark and current purification methods are free from this disadvantage. A detailed review of experimental data on electron attachment in condensed noble gases and purification methods can be found in a recent review by Obodovskiy (2005) and in monographs by Barabash and Bolozdynya (1993) and Schmidt (1997).

Lifetime of electrons and holes in solids

Electron trapping in noble solids is poorly investigated. Increasing electron drift velocity in solids versus liquids may lead to increasing drift path. On the other hand, imperfections in solids may play a role in additional structure traps. Investigators indicated that in condensed krypton at 7 kV/cm drift field (Bolozdynya and Stekhanov, 2005) and in condensed xenon in the range of 1–10 kV/cm electric field (Kirilenov and Konovalov, 1981) the drift path, controlled by electro-negative impurities presented at the level of 10^{-6}–10^{-9} O_2 equivalent, increases by a factor of a few at the transition from the liquid to the solid phase. For very pure argon ($2 \cdot 10^{-10}$ O_2 eqv.), the opposite effect was observed, but in the low field range of 20–300V/cm (Aprile, Giboni and Rubbia, 1985).

As mentioned above, the drift mobility of excess holes in noble solids is several orders of magnitude smaller than that of electrons. The hopping mechanism of hole transfer takes place in these media with part-time self-localisation of holes in polaron states. If the count rate of useful events in detectors exceeds the collection time of holes, a positive space charge builds that affects the normal operation of the detector.

Lifetime of ions

As discussed in Chapter 2, ions or trapped electrons can be extracted from some non-polar liquids. This means that such charge carriers can be used for signal generation in emission detectors. It should be noted, however, that the physical possibilities of the amplification of signals based on ion collection are quite limited in comparison to those based on electron collection.

Lifetime of free-drifting ions is limited with recombination and attachment to impurities, as well as to electrodes due to chemical reactions or imaging forces.

In pure non-polar dielectric liquids, negative ions are formed by electron attachment to electronegative impurities such as oxygen, carbon and nitrogen monoxides, halogenated hydrocarbons or polar molecules, such as water.

Positive parent ions can be neutralised by interacting with impurities having a lower ionization potential. Polar impurities can attach to ions becoming massive complex ions.

3.2.5 *Influence of admixes*

There are three large groups of phenomena associated with *mixtures of condensed dielectrics*:
1) mixtures of organic non-polar dielectrics;
2) mixtures of inorganic non-polar dielectrics (nobles); and
3) mixtures of organic and inorganic non-polar dielectrics.

Mixtures of organic non-polar dielectrics have been used for investigation of electron transport properties in such media. In mixtures of low- and high-mobility electron liquids such as n-hexane and neopentane, the mobility of electrons can be described as

$$\mu_{mix} = \left(\mu_{high}\right)^{x_{high}} \left(\mu_{low}\right)^{x_{low}} \tag{3.28}$$

where x_{high} and x_{low} mark the mole fractions of high- and low-mobility ingredients, and μ_{high} and μ_{low} are the electron mobilities, respectively. This equation is valid for relatively small fraction (< 0.5 mole) of low-mobility liquid; at high concentrations, the mobility falls off faster (Schmidt, 1997). For a mixture of two high-mobility liquids, such as liquid methane and ethane, the mobility of electrons in the mixture can be found from the following expression

$$\mu_{mix}^{-1} = \mu_{high}^{-1} + \frac{n_{low}}{const}, \qquad (3.29)$$

where n_{low} is the concentration of ethane with mole fraction up to 0.25. A similar dependence was observed for several dilute solutions of molecular admixtures in heavy noble liquids and tetramethylsilane (Sowada, Schmidt and Bakale, 1977).

Mixtures of light and heavy noble components have been used for the reduction of diffusion and wavelength shifting in scintillation drift chambers and time projection chambers, using scintillation properties of noble media for position-sensitive registration of particles and radiation imaging (see monograph by Aprile *et al.*, 2006 for more details). In liquid argon, the admixture of Xe and Kr decreases the zero-field electron mobility (Borghesani *et al.*, 1997). At high electrics fields, no significant change in the electron drift velocity and mobility has been observed (Shibamura *et al.*, 1975).

It was shown by Yoshino, Sowada and Schmidt (1975) that the addition of the small amount ($\sim 10^{-4}$ relative concentration) of nitrogen led to an increase of the electron drift velocity at high (> 10 kV/cm) applied field strength and to higher (factor ~ 2) value of the saturated drift velocity. Figure 3.8 shows the influence of the addition of N_2 on the electron drift velocity as a function of the electric field strength in heavy noble liquids. For comparison, data for the pure liquids are shown in solid lines. At lower field strength there is no observed effect of the nitrogen solute. An even more dramatic change is observed when complicated hydrocarbons are added to the noble liquids, as shown in Fig. 3.9 for liquid krypton. As seen, the low concentration of methane may increase the drift velocity up to five times. A similar effect was observed for liquid xenon with a $1.9 \cdot 10^{20}$ cm^{-3} butane concentration added. Low solute concentrations influence the drift velocity at higher field strengths only, while in more concentrated solutions the drift velocity is affected (reduced) in the range of $F < 100$V/cm.

More details on electron transport properties of mixed noble liquids can be found in Obodovsky's review (2005).

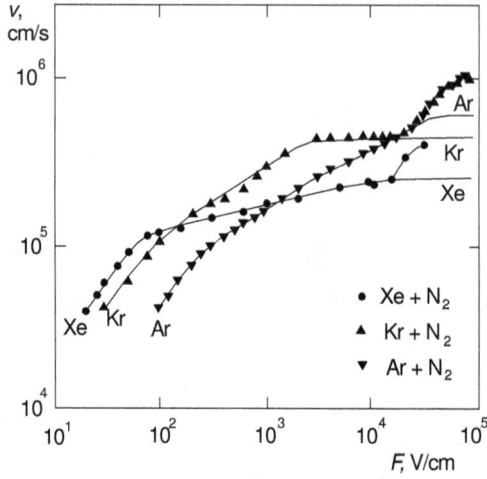

Figure 3.8 Electron drift velocity versus applied electric field in high-mobility liquids (solid lines) mixed with ~ 0.1% of nitrogen and measured at 87 K (Ar), 120 K (Kr) and 165 K (Xe). Redrawn from Yoshino, Sowada, and Schmidt, 1976.

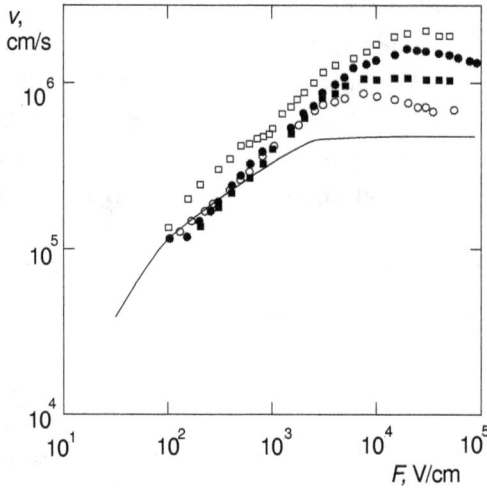

Figure 3.9 Electron drift velocity versus applied electric field in pure liquid krypton (solid line) and in mixes with $5 \cdot 10^{20}$ cm^{-3} methane (open squares), $2.5 \cdot 10^{20}$ cm^{-3} ethane (open circles), $1.2 \cdot 10^{20}$ cm^{-3} butane (closed circles) and $4.7 \cdot 10^{20}$ cm^{-3} butane (closed squares). Redrawn from Yoshino, Sowada, and Schmidt, 1976).

Admixtures of molecular solutes in condensed noble gases effectively reduce the average energy of drifting electrons leading to increased drift velocity, reduced diffusion, and quenching gas gain in equilibrium gas phase due to the absorption of UV light by complicated molecules. In terms of the Cohen-Lekner transport model, the drift velocity in high electron mobility liquids is determined by the relative probability of the elastic and inelastic energy loss rate. Molecular solutes effectively increase the rate of inelastic energy loss rate. This immediately affects the probability of hot emission from condensed nobles. For example, adding 12 mole% CH_4 to liquid krypton increases the emission threshold by a factor of 6 (Bolozdynya, 1991).

3.3 Transfer of charge carriers at interfaces

In Chapter 2, we considered physical conditions for effective electron emission. In this section we focus attention to practical aspects of electron collection in the vicinity of the liquid interface that can affect detection properties of emission detectors.

As we have shown there is a probability that some fraction of the electrons approaching the surface can immediately escape into the rarefied phase, while the rest may be delayed under the surface, change their energy distribution, or become trapped. In heavy noble liquids and solids electrons collected by electric fields can be heated. Hot electrons with energies exceeding the emission threshold and momentum directed to the interface will escape immediately demonstrating a 'fast' component of the emission. The rest of the electrons staying under the surface still have a chance to be emitted as a thermally activated 'slow' component of the emission. Normally, one of the mechanisms dominates. In the case of liquid argon, both components have been observed (Guschin, Kruglov and Obodovsky, 1982b).

The delayed electrons localised under the interface of the dielectric of potential $V_0 < 0$, which are 'waiting' for thermally activated escape, can still be heated by the electric field applied along the interface. Anisimov *et al.* (1984) demonstrated this effect in liquid krypton using the experimental set-up shown in Fig. 3.10. The electrons were generated in LKr near the cathode using short X-ray pulses injected into the two-electrode ionization chamber through the thin aluminium window (Fig. 3.10, left).

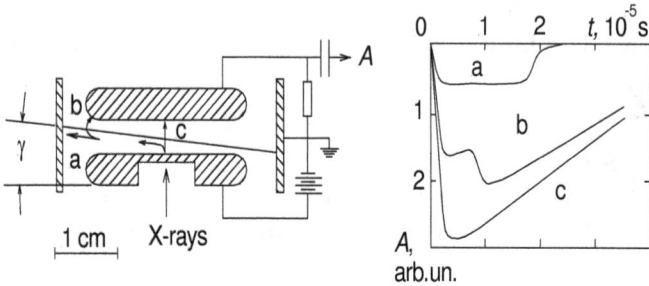

Figure 3.10 Experimental set-up for measurement of electron transfer along and across LKr/vapour interface (left) and typical voltage waveforms observed (right). Redrawn from Anisimov, Bolozdynya and Stekhanov, 1984.

Waveforms of voltage pulses initiated by X-rays have been recorded at different strengths of the applied electric field (Fig. 3.10, right). At electric fields of $F = 100$–1000V/cm, the signals have 10–30 µsec duration and trapezoidal shape (a). At elevated electric fields the waveform gradually transformed from (a) to (b) and (c). The observation was explained by different drift paths for electrons delayed under the interface (see suitable trajectories in Fig. 3.10, left). Since the ionization chamber was tilted by a small angle $\gamma \approx 3°$, a weak component of the electric field γF appeared along the interface and enforced the electrons to drift as shown by arrows in Fig. 3.10, left. At high enough electric fields, either a) the electrons approached the edge of the anode and continued drift to the grounded wall, or b) the electrons, being heated by the increased electric field near the edge of the anode, were emitted as hot electrons. When the electric field achieved the value F_0, electrons emitted as soon as they approached the interface-trajectory and waveform of type (c). The under-interface electron drift velocity has been measured to be the same as the drift velocity in the bulk LKr (Fig. 3.11). These observations give clear evidence that electrons can *drift under the interface* in the quasi-free state.

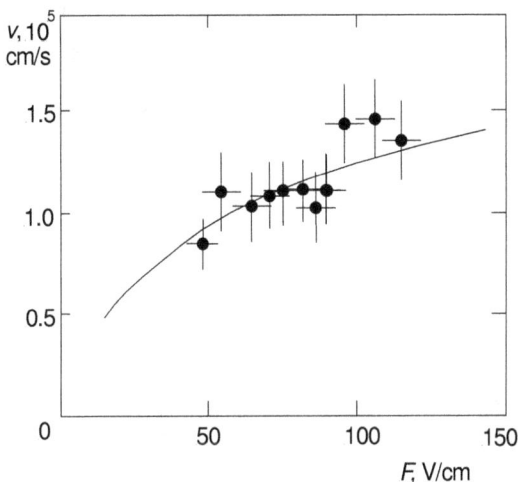

Figure 3.11 Electron drift velocity along the interface measured in LKr at 125 K and in the bulk sample of LKr as shown in solid line. Redrawn from Anisimov, Bolozdynya and Stekhanov, 1984.

In practice, this effect can be used for two applications: 1) for 'cleaning' the interface of non-emitted electrons and 2) for transforming the three-dimensional distribution of electrons originated in bulk working medium of the emission detector into the one-dimensional distribution on the interface before emission occurred (Anisimov, Bolozdynya and Stekhanov, 1984).

3.4 Detection of photons

In emission detectors using noble working media, the detection of photons is important for the completion of two tasks. First, for triggering *time-projection chambers* (TPC) using condensed noble liquids and solids filled with scintillating working media; second, for the detection of ionization signals amplified via the electroluminescence of noble gases. Both processes lead to photon emission in the vacuum ultra-violet (VUV) range. This requires very special methods for efficient collection and detection of the VUV light.

Saturated liquid hydrocarbons are transparent in the visible band of the electromagnetic emission spectrum and effectively absorb the both UV and IR photons. However, no significant emission of photons in the visible range has been observed from condensed methane and more complicated saturated hydrocarbons, such as iso-octane, even with the addition of noble gases.

In this section, we review the most important photon transmission properties of condensed noble media that are effective scintillation and electroluminescence media.

3.4.1 Collection

The propagation of self-scintillation emission through bulk noble liquids and solids have been investigated in the course of developing of fast and effective electromagnetic calorimeters for high energy physics applications (see Barkov *et al.*, 1996; Baldini *et al.*, 2005; Aprile, Bolotnikov, Bolozdynya and Doke, 2006).

Refraction

In most practical cases, non-polar dielectrics used in emission detectors demonstrate high optical *transparency* in the optical band. Collection of ionization electrons in high electric fields is often accompanied with excitation of the media and photon emission. Optical signals may be used to trigger the detector and to measure deposited energy. Knowledge of refractive indices is needed to design effective light collection systems.

The *refractive index* or *index of refraction* of an optical medium is a measure of how the speed of light is reduced inside the medium relative to the vacuum. The refractive index n may be defined via the *dielectric constant* or the relative *permittivity* ε and the relative *permeability* μ as follows

$$n = \sqrt{\varepsilon\mu} \qquad (3.30)$$

For non-polar dielectrics used as emission detector media, μ is close to 1 at optical frequencies and formula (Eq. 3.30) appears in most traditional form: $n = \sqrt{\varepsilon}$. In most cases, $n > 1$, but it can be < 1, as with very high energy photons, such as X-rays.

The dielectric constant of a homogeneous non-polar medium is related to the molecular *polarisability* α and the medium density ρ by the *Lorentz–Lorenz function*

$$\frac{\varepsilon - 1}{(\varepsilon + 2)\rho} = \frac{4\pi N_A \alpha}{3M} = f_{LL} \qquad (3.31)$$

where M is the molecular weight and N_A is Avogadro's number. The Lorentz–Lorenz function adequately describes the dielectric behaviour of condensed noble gases if the polarisability α is properly modified to take account for density (Sinnok and Smith, 1969). Experimentally obtained indices of refraction for popular working media of emission detectors are collected in Table 3.8 and are shown in Fig. 3.12 for liquid xenon as a function of wavelength.

Absorption

Noble gases condense to form colourless liquids and freeze under suitable conditions to form transparent solids. *Self-absorption* of the scintillation light in noble liquids and solids is negligible. Basic absorption properties of condensed noble media can be found in Baldini (1962).

Table 3.8 Indices of refraction for some popular working media of emission detectors.

Medium	Phase	T, K	Wave Length, nm	Index of Refraction	Reference
Ar	liquid	t.p.	546.1	1.2316	Sinnok and Smith, 1969
Ar	solid	t.p.	546.1	1.2334	ibid.
Kr	liquid	t.p.	546.1	1.298	ibid.
Kr	solid	t.p.	546.1	1.346	ibid.
Xe	liquid	t.p.	546.1	1.385	ibid.
Xe	liquid	t.p.	180	1.57	Barkov *et al.*, 1996
Xe	liquid	170	178	1.69±0.02	Solovov *et al.*, 2004
Xe	solid	t.p.	546.1	1.443	Sinnok and Smith, 1969
CH_4	liquid	94	670	1.2872	Badoz *et al.*, 1992
iso-octane	liquid	293	D	1.3914	ChemBlink
n-hexane	liquid	293	D	1.3749	ibid.
TMP	liquid	293	D	1.3347	ibid.
TMSi	liquid	293	D	1.3585	ibid.

Notes: (D) Na D-line (589.59 nm), (TMP) thetramethylpentane, (t.p.) triple point, iso-octane – 2, 2, 4-trimethylpentane, (ChemBlink) Online Database of Chemicals.

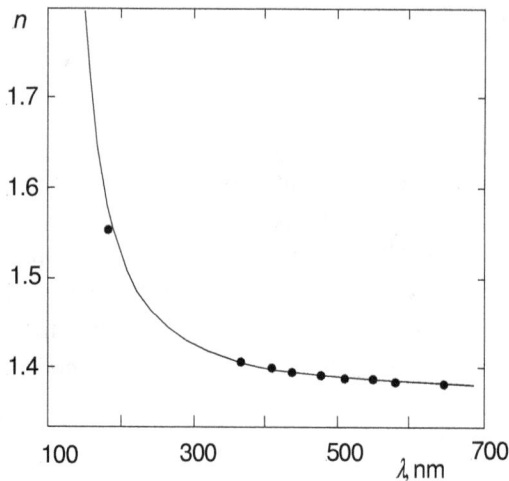

Figure 3.12 Refractive index versus wavelength in liquid xenon near triple point: line — calculations, points — experimental data. Redrawn from Sinnok and Smith, 1969 and Barkov *et al.*, 1996.

The presence of dissolved molecular impurities in noble liquids and solids is a major source of photon absorption from the molecular continuum of the excitation spectra. It was recently shown that liquid xenon effectively dissolves even solid organics (Bolozdynya *et al.*, 2008). However, the admixture of water at *ppm* level (10^{-6} or *part per million*) is the most probable transparency limiter of liquid xenon (Baldini *et al.*, 2005) and, probably, of other noble liquids. Industrial high-purity xenon usually demonstrates an attenuation length of the scintillation light L_a < 1 cm. The attenuation length in massive samples of LXe can be drastically improved (L_a > 1 m) by continuous circulation of the gas through a hot metal getter.

Scattering

There are two effects that should be accounted for in order to reduce scattering in noble liquids. First, there is significant dependence of the refraction index (density) of the liquid on temperature. To reduce this effect, measures should be taken to reduce temperature gradients below 0.1 K/cm. Second, the wavelength of scintillation VUV photons is comparable to the size of natural density fluctuations in noble liquids. According to the formula derived first by Einstein in 1910 (Landau and Lifshitz, 1984), the attenuation of the light due to *Rayleigh scattering*, in which the frequency of light ω is essentially unchanged, goes as

$$L_S \sim \omega^{-4} \tag{3.32}$$

It was experimentally found that this effect is competitive to practically achievable attenuation from residue impurities in noble liquids (Seidel, Lanou and Yao, 2002). The effect can be totally eliminated if the original emission spectrum can be shifted to a longer wavelength (say, to a visible range). Even small wave shifting to 170 nm wavelength of Xe used as a wavelength shifter in 1–3% admixture to LKr and LAr significantly reduces the scattering (Ishida *et al.*, 1997) and improves light collection (Akimov *et al.*, 1993).

The inverse of the Rayleigh scattering length can be written in the form

$$L_S^{-1} = \left(\frac{\omega^4}{6\pi c^4} \right) \left[k_B T \rho^2 k_T \left(\frac{\partial \varepsilon}{\partial \rho} \right)_T^2 + \frac{k_B T^2}{\rho c_v} \left(\frac{\partial \varepsilon}{\partial T} \right)_\rho^2 \right] \tag{3.33}$$

where ω is the angular frequency of the radiation, c is the velocity of light, k_B is Boltzmann's constant, T is the temperature, ρ is the liquid density, k_T is the isothermal compressibility, c_v is the heat capacity at constant volume, ε is the dielectric constant. The theory confirms that the presence of gradients of density and dielectric constant increases the scattering and in some radical cases may cause reduced transparency of the samples.

Rayleigh scattering may be effectively suppressed if waveshifting techniques are used. For example, adding $\sim 0.1\%$ Xe to LKr or LAr shifts the emission spectra to a higher wavelength and significantly reduces the scattering (Ishida *et al.*, 1997). Seidel (Seidel *et al.*, 2002) also pointed out that one could add fluorine (or may be just using Teflon in contact) to LXe to wavelength shift the spectrum.

Calculated and measured Rayleigh *scattering lengths* for liquid noble gases are presented in Table 3.9. The major uncertainty in the calculations arises from the determination of the dielectric constant; Seidel *et al.* (2002) estimated those uncertainties to be factor of 2 in the case of helium, 40% for neon, 35% for argon, 25% for krypton, and 20% for xenon.

Table 3.9 Calculated Rayleigh scattering length (Seidel *et al.*, 2002) and measured attenuation length for liquefied noble gases as cited.

Liquid	T, K	λ, nm	ε	L_S^{calc}, cm	L_0^{exp}, cm
He^4	4.2	78	1.007	600	$> 10^f$
He^4	0.1	78	1.089	20000	
Ne	t.p.	80	1.52	60	
Ar	t.p.	128	1.90	90	66^a
Kr	t.p.	147	2.27	60	32^b, 82^a, 100^c
Xe	t.p.	174	2.85	30	29^a, 40^d, 50^e

Note: (a) Ishida *et al.*, 1997; (b) Anisimov *et al.*, 1984; (c) Akimov *et al.*, 1993; (d) Braem *et al.*, 1992; (e) Chepel *et al.*, 1994; (f) Simmons and Perkins, 1961; (t.p.) triple point.

Reflection

Effective light-collection of scintillation light is important for time projection chambers and three-dimensional imaging cameras. The scintillation flash of condensed noble gases is often used as a trigger in such detectors. Collection of VUV from bulk and massive detectors is a challenging task, and was first considered in the course of developing of scintillation LXe/LKr calorimeters with a granulated multi-cell reflector structure. We may select a few basic technical solutions used to increase the collection efficiency of scintillation photons.

One of the first technical solutions used the relatively high reflective index for 170 nm light in LXe. Braem *et al.* (1992) constructed an aluminium pyramidal light-collecting cell of 7.4x7.4x60 cm, coated with a 30 nm layer of magnesium fluoride MgF_2. Photo-diodes installed on the largest base side of the pyramid reflector filled with LXe have been used to measure 85% reflectivity at 175 nm. Chen *et al.* (1993) tested UV reflectors with total internal reflections made of Kapton, aluminium, glass and Cu-clad printed circuits vacuum deposited with fresh aluminium layer and a thick (0.6 mm) MgF_2 layer to confirm the reflectivity of LXe scintillation light at the level of 85–90%. They also tested light-collimators installed in front of photo-diodes to provide more uniform response function along the 27cm long light collecting cell. However, the flattened response function in the last case was received by cost of the loss a large amount of the scintillation light (Doke and Mazuda, 1999) that may be still acceptable in high energy physics.

In emission detectors operating as time projection chambers, the drift structure should play a role of an effective VUV light collector and at the same time provide high breakdown strength. The first choice is Teflon or PTFE. This material is chemically inert and is one of very few highly reflective materials in the UV and VUV range. Direct room temperature optical measurement of PTFE at 175 nm (Kadkohda *et al.*, 1999) gives a reflectance of ~ 70% or less, but a growing number of results from liquid Xe detectors show that the reflectance is well above 90%, possibly as high as 98%. (Yamashita, 2004; Chepel *et al.*, 2005). Practically achieved high reflectivity of PTEF is not well understood and may be associated with wavelength-shifting effect or total internal refection along Teflon-LXe boundary.

Reflectors coated with *wavelength shifter* can be used for detection of UV light with photo-detectors with spectral sensitivity too poor for direct detection of short wavelength photons. Aluminised Mylar reflectors partly coated with para-terphenyl have been used for detection of scintillation photons with immersed in LXe and LKr photo-multipliers with glass windows as well as for providing a linear response function from pyramid light-collection cells of 2x2x22x4x4 cm (Akimov, Afonasiev *et al.*, 1993; Akimov, Bolozdynya *et al.* 1995). However, it was found that p-TP is not compatible with long electron lifetime (see next section). TPB coated *VM2000* (3M Worldwide) plastic reflectors (99% reflectivity in visible range) have been developed for use in the WARP LAr emission detectors (Grandi, 2005).

Wavelength shifting

Efficiency of VUV light collection can be improved with solid wave shifters such as p-n-phenyls, sodium salisicate, butyl-butadiene and others. Akimov *et al.* (1993, 1996, 1998) used highly reflective, in the blue-visible range, aluminised Mylar reflectors with vacuum-deposited strips of p-terphenyl to construct a long calorimetric cell with a uniform response function of light collection from LXe, LKr, and their mixtures. This approach can be used

  1) to construct long reflector cells;
  2) to eliminate the effect of Rayleigh scattering;
  3) to use inexpensive glass window photo-multipliers.

But all of these can be achieved only through a sacrifice of total collected light, which is not acceptable for some tasks associated with the detection of low-ionizing particles.

There is one important disadvantage of organic wave shifters. As shown by Bolozdynya, Bradley *et al.* (2007), liquid xenon dissolves p-terphenyl and dissolved p-terphenyl traps electrons, limiting their lifetime at the level of ~ 0.1 μs. The lifetime of electrons is peaked around 1 kV/cm drift field strength and is reduced at elevated electric field strength similar to what has been observed in LKr (Fig. 3.6).

Table 3.10 Properties of wavelength shifters for use in noble gas detectors.

	MW, a.u.	MP, °C	AB, nm	BW, nm	EP, nm	τ, ns	QE(λ), %(nm)	t_e, μs	Ref.
p-3P	230.1	212	< 320	80	330–360	2	> 100 (175)	0.3 (LXe)	a, b
p-4P	306.41	> 300	< 340	80	370	0.8–2			a, f
p-5P	382.50	381				0.8–2			a, f
p-6P	458.59	475			325	0.8–2			a, e, f
TPB	358.47	197	< 320	80	440	2	> 97 (128)	~ 100 (LAr)	c, d
NaS	160.11	200	< 350	100	420	8–10	94–99 (121–254)		d, g, h
DPS	332.44				409				d

Notes: (MW) molecular weight; (MP) melting point; (p-3P) para-terphenyl; (p-4P) para-quaterphenyl; (p-5P) p-quinquephenyl; (p-6P) p-sexiphenyl; (TPB) tetraphenyl butadiene; (NaS) sodium solicylate; (DPS) diphenylstilbene; (AB) absorption band; (BW) emission band width; (EP) emission spectrum peak; (τ) emission decay time; (QE) quantum efficiency; (a) Bolozdynya *et al.*, 2008; (b) Belogurov *et al.*, 1995; (c) Benetti *et al.*, 1992; (d) McKinsey *et al.*, 1997; (e) Balzer *et al.*, 2005; (f) Maeda and Miyazoe, 1997; (g) Herb and Van Sciver, 1965; (h) Samson, 1967.

Using heavier members of the *p-n*-phenyl-family (Table 3.10), the electron lifetime can be improved. P-n-phenyls have already demonstrated high chemical stability and high luminescence quantum yield in the blue range (Balzer *et al.*, 2005); and — what is technically convenient and supports high purity — all of them can be deposited by evaporation with elevated sublimation temperature for heavier members. These organic substances attract attention because they are fast, easy to use, and their good performance has been demonstrated in organic light-emitting diodes, organic field-effect transistors, and solar cells (Hu *et al.*, 2005).

One can guess that there are many other molecular substances that can be used as solid wavelength shifters. For example, there is indication that PTFE itself may be working as a weak wavelength shifter with an excitation band in the range of < 290 nm and emission band in the wavelength range of 310–350 nm (Kadkhoda *et al.*, 1998). This effect may explain why PTFE reflectors have been successfully used in older LXe scintillation detectors equipped with photo-multipliers with relatively low quantum efficiency below 180 nm (see, for example, Barabanov *et al.*, 1986) and why modern light collection systems, such as used in XMASS and XENON-10 detectors, have such highly effective light collections (Yamashita, 2004). New advanced optical materials such as Spectralon™ demonstrates similar properties, however, the manufacturer LabSphere does not recommend usage of this reflector in the range of < 250 nm.

In emission detectors filled with liquid argon, *tetraphenyl butadiene* (TPB) wavelength shifter (maximum emission intensity at 438 nm wavelength) has been used to detect 128 nm photons with an *Electron Tubes 9357FLA* photomultiplier equipped with a borosilicate glass window coated with a 200 μg/cm^2 layer of TPB (Benetti *et al.*, 1992). This method allows detecting 128 nm photons with quantum efficiency 9%. The conversion efficiency of TPB was defined to be in order of 80% for wavelength down to 160 nm (Lally, 1994).

Inorganic phosphors such as CaWO$_4$, CdS, ZnS, BaF$_2$ also can be used as wavelength shifters. Advantages of inorganic wave shifters are their inherent high purity, negligible solubility in organics and LXe, low vapour pressure and elevated quantum efficiency at reduced temperatures. However, such substances demonstrate slow decay times (~ 1 μs) and the extremely high temperatures necessary for vacuum deposition make deposition difficult on plastic materials, input windows of semiconductor photo-diodes, and photo-multipliers. For these reasons, inorganic wave shifters have not found broad application in emission detectors technology.

3.4.2 Photo-detectors

Photon detection is one of several techniques often employed for radiation detection. *Photo-detectors* capable of single photon detection, along with scintillating media capable of effectively transferring absorbed energy into photon emission, opened a new era of digital detection devices that revolutionized experimental physics in the latter half of the past century. Fast scintillation signals originating at the moment of particle interaction with the detector medium are used to trigger liquid noble emission detectors and for precision determination of the depth coordinate of the interaction in the time projection technique. Detection of excitation photons generated during electron drift through the gas phase is used to measure deposited energy and position in the plane perpendicular to the drift path.

Photo-electron multipliers

Photo-electron multiplier tubes (PMT) with multi-stage amplification of photo-electron signal invented by L.A. Kubetsky in 1930 and re-invented by V. Zworykin in 1936 (see Lubsandorzhiev, 2006, and references therein) are the most traditional photo-detectors used in many experiments involving generation of scintillation and electroluminescence signals. PMTs are fast, have low noise, and can detect single photons due to photo-electron multiplication factors up to $\sim 10^6$. The successfully run LXe emission detector XENON-10 used 89 1"-square Hamamatsu R8520-06-AL PMTs (one of two PMT arrays is shown in Fig. 3.13) triggered by LXe scintillation to measure energy depositions from particle interactions via electroluminescence generated in the gas phase. ZEPLIN III detector used 31 2" diameter ETL D730/9829Q photo-multiplier tubes for the same purposes. Parameters of PMTs recently used in successful emission detectors are listed in Table 3.11.

There are two requirements for PMTs operating in the vicinity of or inside noble liquids: low operating temperatures and detection sensitivity to extremely short wavelength photons. Low temperatures affect photo-cathode conductive properties. The short wavelength of detected photons limits the selection of PMT input windows to special optical materials (MgF_2 for detection of excitations in argon peaked at 128 nm, quartz for detection of excitations in xenon peaked at 175 nm).

Table 3.11 Photo-multipliers used in noble gas emission detectors.

Photomultiplier	Diameter/ /Length, mm	Photo- Cathode/ /Window	TR, nm	TTS ns	QE(λ), % (nm)	R, mBq	P_{max}, bar
R8520-06-AL	25x25x35	RbCs/Q		~ 1	20(178)	< 1	5
9829Q	50/100	D730/Q		~ 1	18(175)		
R8778	57/112	Bialk/S	160–650	4	26(175)	50	5
R6041Q	57/32	RbScSb/Q		0.75	6(175)		3
9357FLA	203/293	$K_2CsSbPt/G$	300.5 max	~ 4	20(blue)		

Notes: (TR) optical transmittance range; (R) radioactivity; (TTS) electron transition time spread; (S) synthetic silica; (Q) quartz; (G) borosilicate glass.

Figure 3.13 Top array of 48 Hamamatsu R8520-06-AL photo-multipliers in XENON-10 detector.

In first tests of bialkali photo-cathodes, an abrupt decrease in quantum efficiency was observed below certain temperatures (Ichige *et al.*, 1993; Araujo *et al.*, 1997). More detail investigation showed that when a PMT with Sb-Cs or Sb-Na-K-Cs photo-cathode is cooled down to ~ 250 K, the average dark current initially decreases in accordance to the RLD equation (2.12) for thermoionic emission; at lower temperatures the dark current increases with falling temperatures down to 4 K (Meyer, 2008).

Multi-alkali photo-cathodes such as Cs_3Sb and CsK_2Sb were found to be more stable but their quantum efficiency reduces with decreased temperature. To suppress the temperature dependence, the photo-cathode could be evaporated onto a thin conductive layer ($\approx$ 50Å of Pt), or thin conductive strips deposited cross-wise over the photo-cathode area (Fig. 3.16). The Hamamatsu R8520-06-AL photo-multipliers used in XENON-10 detector were designed in this fashion (Fig. 3.13).

UV-transparent input windows are usually fragile and, in case of quartz, can be easily penetrated by helium molecules diluted in air which then degrade the vacuum inside the tube. To avoid these dangers special precautions are required when operating such PMTs.

All photo-multipliers exhibit an *after pulsing* phenomenon — 1–10 photo-electron pulses following the physical multi-photo-electron signal (Wright, 1999). This effect is associated with the ionization of residue gas inside the PMT and may be used to diagnose leaking PMTs. The effect can be effectively suppressed in PMT arrays operating in coincidence.

Solid state photo-detectors

Solid state photo-detectors attract attention due to inherent high mechanical rigidness, capability to operate in strong magnetic fields and at elevated pressures. The main disadvantage of semiconductor photo-detectors is relatively high noise. This disadvantage becomes less significant at reduced operating temperatures specific for cryogenic detectors employing condensed noble gases as working media.

In the 1990s, much effort was devoted to the development of noble liquid scintillation calorimeters to measure the energy of relativistic particles in high energy physics accelerator experiments (Masuda *et al.*, 1991; Bream *et al.*, 1992; Chen *et al.*, 1993). The surface barrier *silicone photo-diode* (SiPD) was one of the first photo-detectors tested for the detection of scintillation light in liquid xenon (Masuda *et al.*, 1991) with quantum efficiency 22% at 175 nm. To maximise the transmission of the input window of the SiPD, the facing electrode was made of a gold mesh with 20 μm x 100 μm strips and 200 Å thicknesses. Afonasiev *et al.* (1992) demonstrated another approach using a p-n-junction open to the input surface. A 5x5 mm^2 SiPD made of 0.1 mm thick p-Si was coated with n-Si mesh with pitch of 50 μm. The input surface was protected with half-wave UV transparent SiO$_2$ protection layer. The dark current of the SiPD immersed in LKr at 120 K was less than 1 nm, a capacitance was 50 50 pF at 10V bias. With this SiPD scintillations of LKr irradiated with α(^{239}Pu)-source placed at 2 mm distance from the input window was measured with energy resolution 16% FWHM.

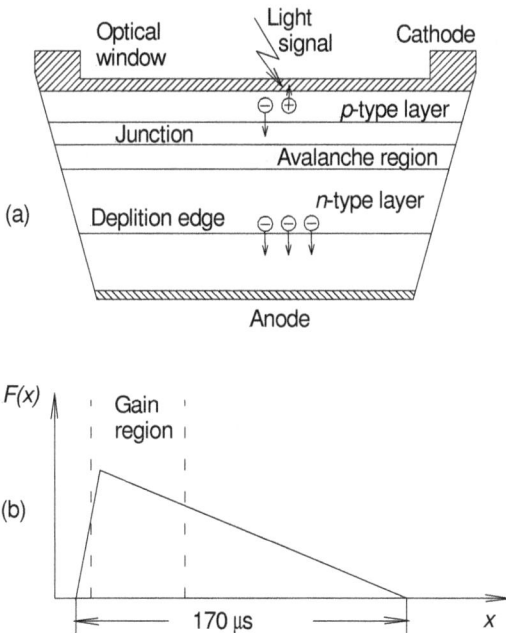

Figure 3.14 Schematic cross-section of the LAAPD (a) and electric field profile (b). Redrawn from Moszynski *et al.*, 2002.

Recent progress in semiconductor technology has led to the development of a very effective silicon photo-diode with internal amplification of the electronic signal (Sadygov *et al.*, 1996). One of the best commercial devices of this type is Large Area Avalanche Photo-diode (*LAAPD*) with a 16 mm diameter input optical window developed by Advanced Photonics (Boisvert *et al.*, 1996). The LAAPD cross-section is shown in Fig. 3.14. Photo-carriers from photon absorption are generated in the undepleted region of 5–8 μm thickness under the optical window. These carriers then enter the drift region and are accelerated to energies sufficient for ionization and the electron avalanche develops here. The *avalanche photo-diode* has a typical gain of 100–300 and may be increased at reduced temperatures.

It was shown that the LAAPD can operate inside a high-pressure He^3 scintillation detector without gas purity degradation (Arodzero *et al.*, 2004) and inside liquid xenon (Solovov *et al.*, 2002; Ni *et al.*, 2005).

Silicon photo-multiplier (SiPM) is a miniature array of APDs, each 20 to 100 μm in size and highly packed with densities up to 1000 APDs per square millimetre at the silicon wafer (Buzhan *et al.*, 2001; 2003). Each SiPM pixel operates in *Geiger counter mode* at a bias voltage exceeding the breakdown voltage by 10–20% and is individually coupled to integrated quench electronics. Due to fast developing space charge, the Geiger discharge is stopped after < 1ns of the electron avalanche. At typical pixel capacitance ~ 100 fF and a few applied volts, each pixel accumulates and electric chare of 100–1000 pC that responds to an electron multiplication factor of up to 10^6, i.e. the same order as vacuum PMT gain and good enough to detect a single photon. Since all SiPM pixels are loaded to the same impedance, the output signal is a sum of the signals from all pixels fired or is proportional to the number of detected photons.

A SiPM produced by Pulsar Enterprise (Moscow) was tested by Aprile, Cushman *et al.* (2005) using the scintillation of LXe irradiated with ^{241}Am α-source. They obtained a gain of $1.8 \cdot 10^6$ with a 52V bias voltage, single photon sensitivity and 22% quantum efficiency at 5.5% detection efficiency for 178 nm LXe scintillation photons. One important advantage of the SiPM is a linear dependence of its gain to applied voltage in contrast to the exponential dependence specific to PMTs. The main disadvantage of the SiPM is a relatively high cost per unit of sensitive area.

Hybrid photo-detectors

Hybrid photo-detectors (HPD) combine advantages of photo-multipliers (high gain) and solid state detectors (linearity). The HPD consists of vacuum tube with a photo-cathode and semiconductor diode, or other electron detectors, installed instead of multi-stage electrode electron multiplying structure (Viatskin and Makhov, 1958; Kalibjian, 1965). The photo-electrons escaping from the photo-cathode are accelerated by an electric potential up to 20 kV and impact the solid state detector. The gain obtained is proportional to the number of secondary carriers generated in the solid state detector and to the photo-electron-accelerating voltage. The HPD with Si-diode detector has demonstrated excellent single photo-electron sensitivity and more than four orders of magnitude of linearity (DeSalvo *et al.*, 1992).

A multi-channel plate can be used in hybrid photo-detectors to detect accelerated photo-electrons. This kind of photo-detector is called a proximity focussed MCP photo-multiplier and is sometimes used as a fast position-sensitive photo-multiplier. One of the first devices like this was the Hamamtsu R1564X (Hayashi, 1982). The multi-channel plate can be used as an UV-sensitive photo-target in video camera tubes (Ammosov *et al.*, 1986). The device has been tested for imaging a ring-like profile of a nitrogen laser beam (333 nm wavelength, 2 mm diameter) and has demonstrated $< 0.1 mm^2$ two-dimensional position resolution.

Recently, the idea of hybrid photo-detectors received new life. For next generation dark matter experiments using multi-tonne noble liquid detectors, it was proposed to develop a relatively inexpensive large-area photo-detector with a wide dynamic range. One idea is to equip a *QUPID* with a semiconductor diode, as shown in Fig. 3.15. In another idea, accelerated electrons are detected with a thin scintillator decoupled to the optical fibre connected to the photo-diode placed out of the vacuum tube.

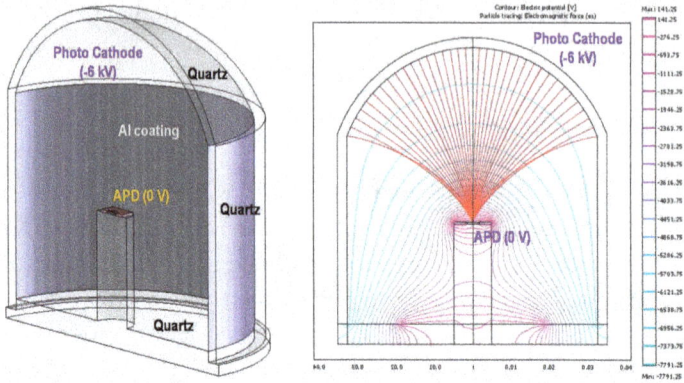

Figure 3.15 Schematic drawing of QUPID (left) and electrical field distribution inside the vacuum tube (right). Courtesy of K. Arisaka.

Vacuum photo-tubes are basic photo-sensors used in upcoming dark matter and solar neutrino search experiments (Fig. 3.16).

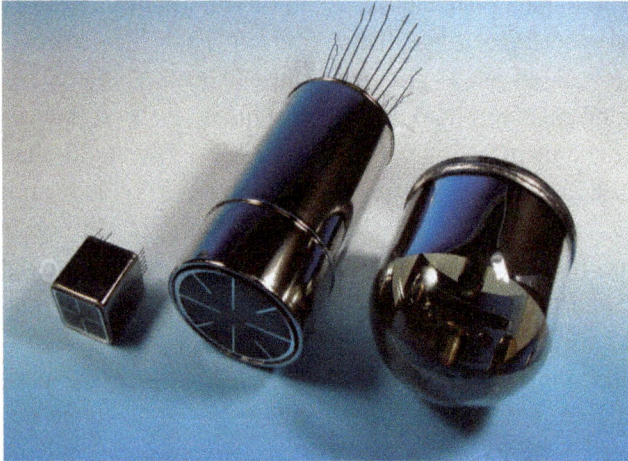

Figure 3.16 Photograph of low-radioactive Hamamatsu photo-multipliers for cryogenic operations (left to right): 2.5x2.5 cm^2 R8520, 5 cm dia. R8778, 7.5 cm dia. QUPID. Courtesy of K. Arisaka and Hamamatsu Photonics Co., Japan.

Open photo cathodes

Open photo-cathodes such as CsI and CsBr can be used for effective registration of UV emission in noble gases. It was shown that CsI can provide a high quantum efficiency of 20–30% range of 120–200 nm for the detection of scintillation photons from krypton and xenon (Rabus *et al.*, 1988). The CsI coating may enhance the performance of gas electron multipliers (Bondar, Buzulutskov *et al.*, 2007). Time stability of open photo-cathodes and background from direct interactions is an open question that needs special consideration.

Placing an open photo-cathode in the noble liquid with a negative electron ground state V_0 reduces the electron escape barrier into the liquids and increases the quantum efficiency of photon detection. Aprile *et al.* (1994) tested CsI photo-cathodes (CsI work function $\varphi = 6.1$ eV) for scintillation detection in condensed krypton and xenon. High quantum efficiencies of 60% in solid Xe, 31% in liquid Xe, 17% in liquid Kr, and 10% in liquid Ar have been achieved with 50-nm thick CsI photo-cathodes illuminated with a 220 nm laser beam. Excellent intrinsic energy resolutions of 2.8% FWHM for liquid Xe, 5.2% FWHM for liquid Kr, and 8.4% FWHM for liquid Ar have been measured for scintillations generated by 5.5 MeV alpha-particles.

Photo-emission could be effectively enhanced with high electric fields (up to 100 kV/cm) applied to the thin films of CsI via the Schottky effect. Boutboul *et al.* (1999) showed that the electron affinity is reduced in CsI from 0.2 to 0.01 eV at electric field strength of $5 \cdot 10^5$ V/cm. High field strength can be arranged around the wire cathodes (Buzulutskov *et al.*, 1995).

General properties of different photo-sensors for emission detectors are compared in Table 3.12. Photo-multipliers are still most often used in large emission detectors due to favourable combination of photo-detection and commercial properties. Both vacuum and solid state photo-sensors represent 'pure' technique compatible with long electron lifetimes in liquid xenon. Open photo-cathodes have demonstrated good operation properties in LAr emission detectors; however, there is still a concern about achievable purity of LXe working in contact with CsI.

Table 3.12 Comparison of properties of photo-detectors used in noble gas emission detectors.

	PMT	APD	HPD	SiPM	CsI/GEM	CsI in Liquids
QE(λ),%(nm)	25(178)	50(175) 80(400)	25(178)	50(175)	20(178)	31(220) in LXe
Gain	10^6-10^7	100-300	10^3	10^6	10^4	1
Area, cm^2	~ 100	~ 1	~ 100	0.1	~ 100	~ 100
Voltage, kV	1-2	0.1-1	~ 10	0.025		
Threshold, phe.	~ 1	~ 10	~ 1	~ 1	~ 1	
Timing, ns	0.1-1	1-10	0.1	~ 0.03		
Position Sensitivity, mm	~ 10 w/array	~ 5 w/array	~ 30 w/array	~ 1	~ 1	
Dynamic Range	~ 10^6			~ 10^3/mm^2		
Price, k\$/cm^2	30	500	10	1000	0.3	0.1

Notes: (PMT) photo-electron multiplier tube, (APD) avalanche photo-diode, (QE) quantum efficiency; (HPD) hybrid photo detector.

3.5 Amplification of signals in rarefied phases

An important advantage of emission detection technology is the possibility to amplify signals in physical processes such as electron multiplication or generation of intensive light emission with very few electrons drifting in electric fields of minimally high strengths. Let us consider these processes.

3.5.1 Electron multiplication

In the presence of electric fields, electrons drifting in a gas can gain a sufficient amount of energy from the field between two collisions to cause ionization or excitation of atoms. The secondary electrons can also gain energy for next ionization. This process can lead to avalanche multiplication of electrons or *gas amplification*.

If drifting electron can ionize α atoms per unit length, increasing number of electrons is calculated from equation $dN = \alpha N_0 dx$ to be

$$N(x) = N_0 \exp(\alpha x), \qquad (3.34)$$

if α does not depend on N_o, the initial electron concentration. The coefficient α is called the *first Townsend coefficient* and depends on the field strength, ionization potential of the media, and the cross-section of electron–molecule interactions.

Since electron multiplication accompanies excitation of the media, the influence of photon emission should be taken in account. Intensive photo-ionization of the gas or photo-electron emission from electrodes may initiate a breakdown of the gas. For this reason, the most effective gas media used for electron multiplication usually contain *quenching admixtures* to absorb UV photons, initiating photo-ionization and photo-emission. Quenching admixtures are usually organic gases with complicated molecules.

Emission detectors using non-polar dielectrics as working media can operate without adding special quenching admixtures. For example, a saturated vapour above pure liquid *iso-octane* (2,2,4-trimethylpentane, 100 torr pressure at room temperature) can work as an effective gas multiplication media with gas gain $G \sim 10^4$ (Bolozdynya, Lebedenko *et al.*, 1978). Gas gain may be improved adding noble gas (argon at 80 torr) to an organic vapour (2,2,4,4-tetramethylpentane) as shown by Anderson, Charpak, Holroyd and Lamb (1987).

Gas multiplication in pure noble gases is very unstable. The reduced to the gas pressure Townsend coefficient α/p has a value about one *ion pair per cm drift per torr* at reduced electric field strength $F/p = 100$ V/cm/torr for all pure noble gases. The gas gain (amplification factor) of up to 10^6 is possible only in noble gases doped with quenching molecular impurities. Despite many attempts in pure noble gases, liquids, and solids, only poor gas gain of < 100 has been achieved and no detectors of practical importance have been developed with electron multiplication (see review by Barabash and Bolozdynya, 1993) and recent experiments (Policarpo *et al.*, 1995, and Kim *et al.*, 2004).

In order to overcome instability of gas multiplication in emission detectors with pure noble working media, Dolgoshein *et al.* (1973) considered wire electrodes immersed in liquid argon and heated by electrical current. Boiling argon surrounds the wires as a bubble jacket and gas multiplication with space-limited gas gain in Geiger mode occurred inside the bubbles. Unfortunately, this method did not find practical applications because it requires extremely high operating power (~ 1 W/cm heated wire or ~ 1 MW for 1-ton emission detector).

Anisimov *et al.* (1986) considered the possibility of using methane as a quenching admixture in emission detectors. They constructed a two-electrode emission chamber with a wire anode (20 μm gold-plated tungsten wires with 1 mm pitch) placed in gas phase above ~ 2 mm layer of liquid krypton. A pulsed X-ray tube was used as a source of ionization and methane as a quenching admixture in different concentrations. The best gain ~ 500 was achieved in Kr + 39%CH$_4$ gas mixture. However, the krypton admixture essentially increased the threshold of electron emission (factor 6.4 at 12%CH$_4$) and this approach also has been claimed as impractical.

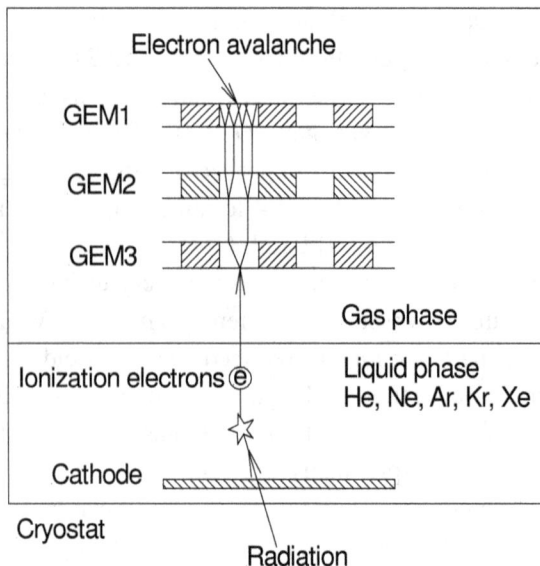

Figure 3.17 Schematic view of an emission detector with GEM electron multiplier. Redrawn from Bondar, Buzulutskov *et al.*, 2006.

The most promising results have been achieved with *gas electron multipliers* (GEM) placed above pure noble liquids in emission detectors (Bondar, Buzulutskov *et al.*, 2004, 2006). GEM foil were produced in

CERN workshop and have the following parameters: 50 μm thick Kapton copper plated foil, 70 or 55 μm hole diameter, 140 μm hole pitch, 28 mm^2 active area. The distances between the first GEM and the chamber bottom and between the GEMs are 5 and 2 mm, respectively as shown in Fig. 3.17.

The triple-GEM structure achieved gas gains exceeding 10^4 in He, Ar, Kr, Xe gases in the temperature range of 120–300 K. However, with triple-GEM placed in the gas phase above heavy noble liquids the best gains measured were 200 in Xe, ~ 10^3 in Kr, ~ 5·10^3 in Ar (Fig. 3.18). The difference may be understood by taking into account that the vapour phase is enriched with polymer noble gas clusters A_n having reduced excitation and ionization potentials. Note that adding up to 2% CH$_4$ in xenon did not improve the gas gain.

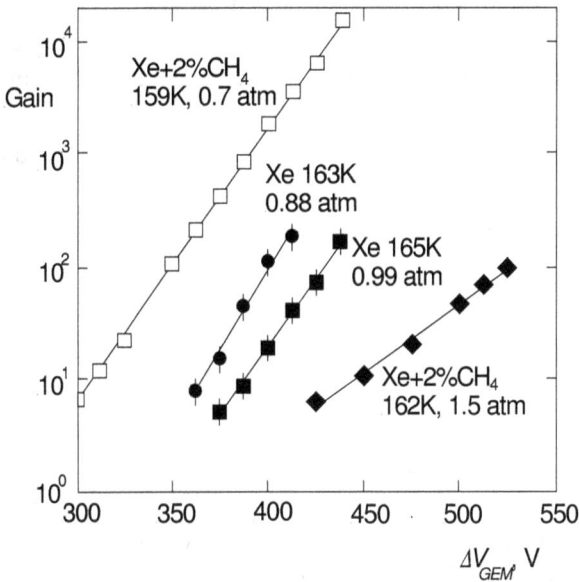

Figure 3.18 Gain-voltage characteristics of the triple-GEM measured in Xe emission detector with pulsed X-ray tube: the cathode is covered with LXe (closed signs) or uncovered with LXe (open signs) at 4.0 kV/cm electric field in LXe. Redrawn from Bondar, Buzulutskov *et al.*, 2006.

3.5.2 *Electroluminescence of noble gases*

If the energies of drifting electrons are slightly below the ionization threshold they effectively excite atoms in collisions and generate *electroluminescence* (called sometimes as *proportional scintillation*) in processes similar to generation of scintillation by high energy particles:

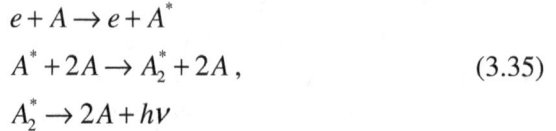

$$e + A \rightarrow e + A^*$$
$$A^* + 2A \rightarrow A_2^* + 2A \,, \qquad (3.35)$$
$$A_2^* \rightarrow 2A + h\nu$$

The second reaction is dominated at gas densities $n > 10^{10}$ cm^{-3}. This is the case of emission detectors employing noble liquids in vicinity of their triple points.

The emission spectra of electroluminescence are similar to those of scintillation (see below Section 3.4.1 for details). There are data showing that some part of the photon emission is released in the infra-red (IR) range (Carugno, 1998). In dense and heavy noble gases such as xenon, an additional mechanism of electroluminescence generation can occur because of significant presence of free *dimers* or *clusters* or even formation of the conduction band

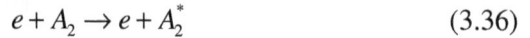

$$e + A_2 \rightarrow e + A_2^* \qquad (3.36)$$

Electroluminescence is a threshold process as shown in Fig. 3.19. The threshold value of reduced electric field value is about the same ($\sim 3 \cdot 10^{-17}$ Vcm^2) in xenon and krypton gas. However, there is a weak photon emission observed below the threshold that is predominantly associated with *bremsstrahlung*.

At uniform electric field the intensity of the *electroluminescence yield Y* , or the number of photons generated by one drifting electron n_{ph}, per unit of the drift path x [cm], for xenon gas at room temperature may be defined via the reduced electric field strength F/p [kV/cm/bar] and the gas pressure p [bar] using the empirical equation (Belogurov *et al.*, 1995)

$$Y = dn_{ph}/dx = 70(F/p - 1.0)p \qquad (3.37)$$

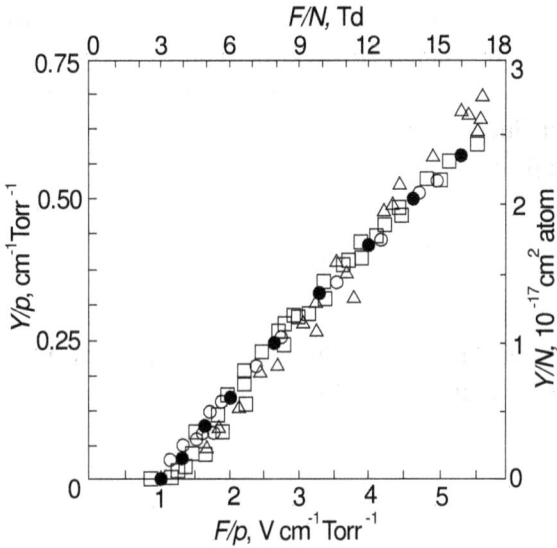

Figure 3.19 Reduced light output of electroluminescence of xenon gas at 293 K temperature and normal pressure in dependence on the reduced electric field strength. Redrawn from Conde, 2004.

The best value of the electroluminescence yield was measured to be 1700 photons/cm at $F/p = 3.4$ kV/cm/bar and 4 bar xenon pressure (Belogurov *et al.*, 1995). Taking into account that the energy of a single photon of 172 nm wavelength is about 8.4 eV, one can calculate the efficiency of converting the energy of the electric field into photon emission: $\xi = (1700 \cdot 8.4)/(3400 \cdot 5) = 84\%$. The rest of the energy acquired by drifting electrons with the electric field is probably converted into IR radiation. Taking into account that at extremely high electric fields in pure xenon the charge multiplication factor can be about 10, we may estimate the conversion efficiency of the electroluminescence process to be $\sim 8\%$.

Electroluminescence has been observed in all noble gases and noble gas mixtures. In mixtures containing $> 0.1\%$ Xe, the light output and electroluminescence spectrum is very similar to that of pure xenon. Note that electroluminescence has been observed in different aggregate states of noble gases but it is practically used only in gases. The process of electroluminescence is linear, in contrast to the avalanche-like charge (electron) multiplication. This is because the energy of drifting electrons is mostly dissipated via the emission of photons, which do not participate further in the process. Due to this circumstance, electroluminescence can provide lower fluctuations and better energy resolution than the gas gain amplification process.

3.6 Detection of signals in cryogenic solids

Transitions of excess electrons between different materials can be as advantageous for detector development as transitions between different phases of the same dielectric (Bolozdynya, 2006). Let us suppose that a single free electron, or a single electron in the conduction band, is created as a result of scattering of an incident particle off a xenon nucleus. The threshold energy for this process is near the band gap ~ E_g. An electric field is applied to collect electrons from the bulk solid xenon (SXe) to a *quasi-particle transition edge sensor* (TES), consisting of a superconductive aluminium fin connected to a superconductive tungsten quasi-particle trap. The ground state of electrons in the conduction band of SXe is about $V_0 = -0.45$ eV relative to the vacuum level. The Fermi level of Al is $E_F = -4.54$ eV. So, the electron falling down into the Al anode deposits $E_F - V_0 \sim 4$ eV energy. The deposited energy may destroy several Cooper pairs (0.36 MeV bonding energy) and generate quasi-particles (QP). The quasi-particles diffuse into the superconductive tungsten (transition temperature < 100 mK) QP-trap and, breaking Cooper pairs there, deposit a heat pulse measured with a SQUID in the scheme described, for example, in Cabrera *et al.*, 1998.

At the moment of interaction, prompt phonons and excitons will be generated. Prompt phonons and scintillation photons (emitted by decaying excitons), as well as electrons, can be measured with the TES. Then, three signals from a single act of particle interaction in SXe can be measured. Comparison of different signals could be used to distinguish useful events from background. Drifting electrons could be 'heated' by the applied electric field in order to amplify the ionization signal and generate additional phonons due to the *Neganov–Trofimov–Luke effect*, which is similar to the electroluminescence in noble gases (see, for example, Trofimov, 1996).

Note that other solid noble gases can be used in this detection scheme providing even larger potential drop across the solid gas–aluminium boundary (for example, $V_o(SNe) - E_F(Al) = 5.6$ eV). However, solid xenon is the most attractive because of its inherent high density, low E_g, and low excitation energies. Unfortunately, properties of solid heavy noble gases at cryogenic temperatures are poorly explored. In particularly, there is no available data on phonon propagation through the bulk crystals and through the boundary with TES. This subject may excite the next generation of researchers.

Chapter 4

Emission Ionization Chambers

Ionization chambers are radiation detectors that collect and measure the charge produced by radiation in a working medium (Rossi and Stoub, 1949; Veksler, Groshev and Isaev, 1949). Emission ionization chambers contain a two-phase working medium. They have been used for investigating the emission properties of non-polar dielectrics, including condensed noble gases, liquid at room temperature saturated with hydrocarbons, condensed mixtures of noble gases and hydrocarbons and superfluid helium (Hutchinson, 1948; Careri, Fasoli and Gaeta, 1960; Bruschi, Maraviglia and Moss, 1966; Surko and Reif, 1968; Schoepe and Rayfield, 1973; Boyle and Dahm, 1976; Boriev *et al.*, 1978; Guschin *et al.*, 1982; Bolozdynya, 1986). A number of diode ionization chambers have been equipped with optical windows used for observing the behaviour of the charged liquid surface (Taylor and McEwan, 1965), deformations of surfaces due to two-dimension electron states (Shikin and Leiderer, 1981) and the extraction of ions through 'geysers' on the surface of superfluid helium (Volodin and Khaikin, 1979). In this chapter, we focus our attention on the most technically original solutions for two-phase emission ionization chambers with electrical signal readout.

4.1 Emission ionization chambers using heavy noble gases

The first emission detector was built by Hutchinson (Hutchinson, 1948) and consisted of a three-electrode parallel-plate ionization chamber with a two-phase solid–gas argon working medium. An anode and a screening Frisch grid were placed in a few millimetres gap of gas above the cathode coated with a thin (~ 1 mm) layer of solid argon.

Electrical pulses were observed when the solid argon was irradiated with gamma rays and the applied electron extraction electrical field exceeded 500V/cm. Hutchinson interpreted the observed effective emission of fast negative charge carriers as ionization electrons.

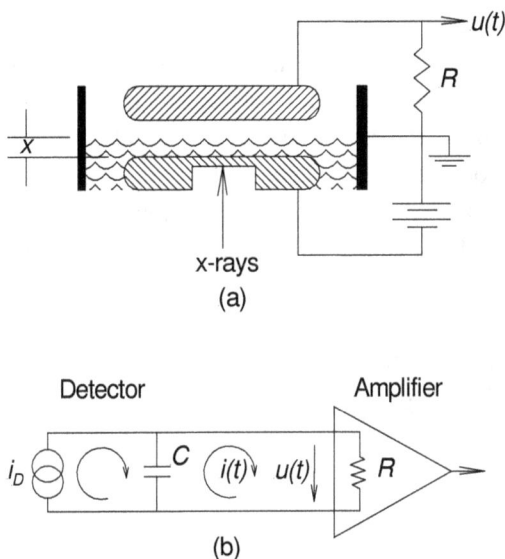

Figure 4.1 Ionization emission chamber (Bolozdynya, 1986) for the investigation of emission processes in liquid and solid krypton irradiated by an X-ray pulsed tube (a) and the equivalent scheme of the charge sensitive preamplifier readout (b).

A miniature two-electrode ionization chamber (Bolozdynya, 1986) with a pulsed X-ray tube is one of the typical emission detectors used in 1980s for the investigation of electron emission from condensed noble gases. The detector consisted of two 25 mm diameter electrodes with a 4 mm gap between them and an 8 mm diameter and 0.4 mm thick aluminium window for the introduction of X-ray radiation through the cathode as shown in Fig. 4.1a. A controlled BSV-7 X-ray tube was used to generate microsecond pulses of X-ray radiation with a maximum energy of 35 keV at 100 Hz rate. The X-ray radiation absorbed in the < 0.1 mm thick layer of the condensed krypton generated a compact (in drift direction) cloud of electrons. The cryogenic system cooling this device served as a prototype for many following developments of condensed noble gas scintillation and emission detectors. The detector was surrounded by a copper thermal shield equipped with heaters and was immersed in the cold nitrogen gas bath surrounded by a liquid nitrogen jacket. The temperature was regulated by the pressure of the nitrogen gas and/or electrical heaters. The system demonstrated exceptional stability (< 0.1 K/h) and it was quite simple in operation. Normally, the ionization chamber was connected to the *charge-sensitive preamplifier* with a discharge time that was much larger than the times of any transition processes in the detector (such as the electron drift time or emission time). The dimensions of the cloud of ionization electrons were assumed to be much smaller than the dimensions of the detector ('point-like ionization').

Upon the ionization of the noble liquid, the two-electrode ionization chamber with an applied voltage V becomes a source of the electric current $i(t)$, charging the capacitor $C = C_D + C_{CSA}$ where C_D is a capacitance of the detector and C_{CSA} is a capacitance of the preamplifier input itself (Fig. 4.1b)

$$i(t) = u / R + C \cdot du / dt \qquad (4.1)$$

A solution to equation (4.1) takes the following form:

$$u(t) = \frac{1}{C} \cdot \exp\left(-\frac{t}{RC}\right) \cdot \int_0^t i(t) \cdot \exp\left(\frac{t}{RC}\right) dt \qquad (4.2)$$

During the collection of ionization electrons of total charge Q_1 through the condensed phase, the electric current in the circuitry is $i_1(t) = Q_1/t_1$ where t_1 is the drift time through the condensed phase (see Fig. 5.4). Then, as the electron cloud heads towards the interface surface, the magnitude of the ionization signal approaches

$$u_1 = \frac{Q_1}{C} \cdot \frac{\tau}{t_1}\left[1 - \exp\left(-\frac{t_1}{\tau}\right)\right] \qquad (4.3)$$

where $\tau = RC$. If some of the electrons are lost during emission, the electric current in the circuitry changes to the value of $i_2(t) = Q_2/t_2$, where

t_2 is the drift time through the gas phase and Q_2 is the charge of extracted electrons. After all the electrons are collected at the anode, the magnitude of the ionization signal is

$$u_2 = u_1 + \frac{Q_2}{C} \cdot \frac{\tau}{t_2} \left[1 - \exp\left(-\frac{t_2}{\tau} \right) \right] \qquad (4.4)$$

By analysing the $u_2(F_2)$ dependence on electric field strength in the condensed phase, the emission threshold can be obtained as shown in Fig. 4.2. The emission coefficient is

$$K = \frac{Q_2}{Q_1} = \frac{u_2 - u_1}{u_1} \cdot \frac{t_2}{t_1} \cdot \frac{1 - \exp(-t_1/\tau)}{1 - \exp(-t_2/\tau)} \qquad (4.5)$$

and can be obtained by examining the waveform like the one shown in Fig. 5.4b.

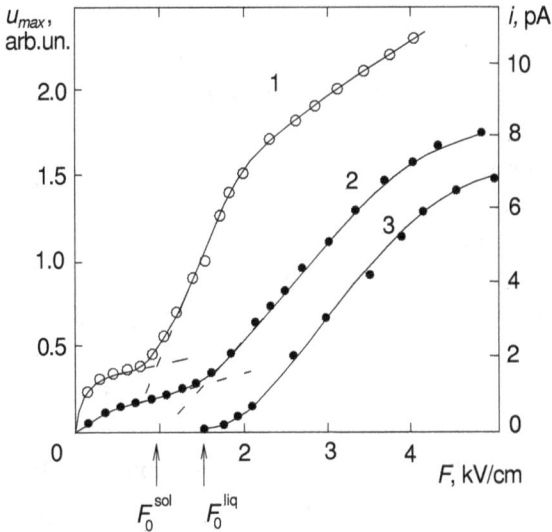

Figure 4.2 Emission curves of the induced voltage pulse u_{max} as measured by a charge sensitive preamplifier (1, 2) and the emission current i as measured by an electrometer (3) as a function of electric field strength in solid (1) and liquid (2, 3) krypton near the triple point. The emission thresholds are indicated by the vertical arrows.

The emission threshold and emission coefficients for liquid and solid krypton and krypton–methane mixtures at different temperatures have been measured using this method (Bolozdynya, 1985, 1986). Data from these measurements have been used to improve our understanding of the origin of quasi-free electron emission from non-polar dielectrics (see Section 2.3.3).

Figure 4.2 shows the typical emission curves of the induced voltage pulse $u_2(F)$ measured by a charge sensitive preamplifier and the emission current $i(F)$ measured with an electrometer as a function of the electric field in condensed krypton (Bolozdynya, 1986). The threshold field of emission has been defined by linearly extrapolating branches of the curves as shown with dashed lines. This emission detector has been used for the observation of electron drift through and along the free surface of liquid and solid krypton and for the investigation of the influence of methane doping on the emission of hot electrons as described in Section 3.6.

The largest emission ionization chamber (41 cm diameter and 6 cm gap with two-electrodes) has been used for measuring the concentration of [85]Kr isotope in a natural mixture of krypton isotopes rectified from air (Anisimov *et al.*, 1989). This experimental study happened around the year of the Chernobyl catastrophe and allowed for the observation of the elevated (for about 100%) concentration of [85]Kr in the product of the Ukrainian Lisichansk plant of rare gases right after the accident. This large ionization chamber has been tested as an emission detector for monitoring the purity of the liquid krypton (see Section 8.8). In this detector, large fluctuations in the emission current were observed in the range of the extraction fields approaching the threshold electric field. The effect had not been observed before with smaller emission detectors. This effect is probably associated with the instability of the liquid surface charged with non-emitted charge carriers. Similar observations have also been reported by others (Yamashita *et al.*, 2005).

4.2 Liquid helium emission ionization chambers

The observation of emission current in an ionization chamber partly filled with liquid helium led to the discovery of tunnelling electrons from localised states in the liquid into the equilibrium gas phase (Careri, Fasoli, Gaeta, 1960; Bruschi, Maraviglia and Moss, 1966).

Using ionization chambers with gated grids, Surko and Reif (1968) discovered a new kind of neutral excitations called *rotons* and observed electron emission associated with these excitations from superfluid helium activated by alpha-particles. Schoepe and Dransfeld (1969) used a rotating (10–100 rpm) cryostat in order to generate *quantum vortices* in superfluid helium and show that vortices can drag electrons through the interface without delay and without overcoming any noticeable barrier. Figure 4.3 shows a schematic drawing of the test cell and the current measured at a collector C versus the temperature at different rotational frequencies. Ionization electrons generated by a radioactive source S drifted between a grid G and an anode A in a direction parallel to the free liquid surface and perpendicular to vortex lines generated by the rotation. The emission current was measured with a collector C placed above the liquid surface. With an accuracy of about 1%, the current was identical to a current measured with the collector fully immersed in the liquid. Without rotation, no current ($< 10^{-14}$A) at C was detected.

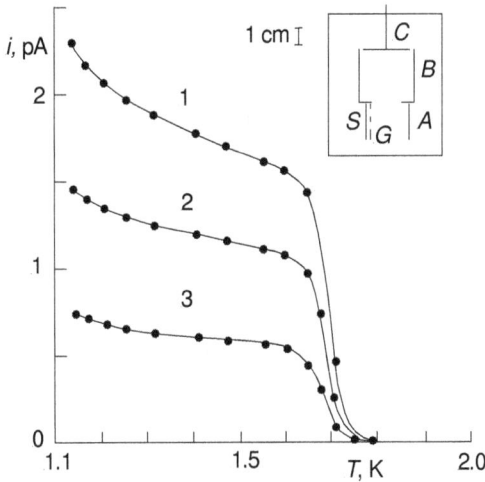

Figure 4.3 Schematic drawing of the rotation cryostat and emission current of electrons dragged by vortices through interface of superfluid helium at 80 min^{-1} (1), 40 min^{-1} (2) and 21 min^{-1} (3) rotation speeds. Redrawn from Schoepe and Dransfeld, 1969.

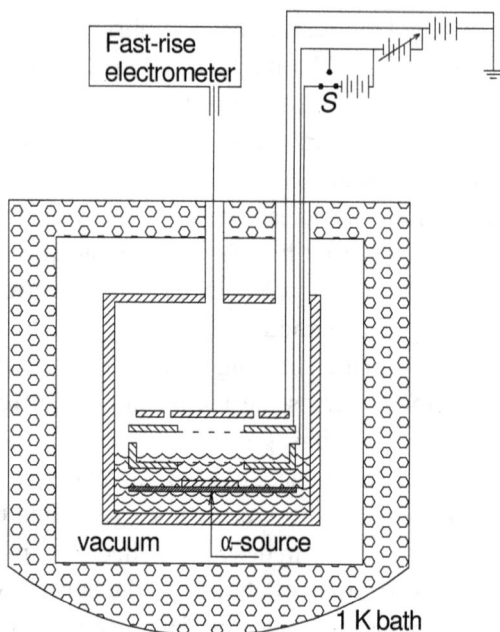

Figure 4.4 Ionization emission chamber with an electrically gated cathode and an anode screened by a Frisch grid for investigating the emission process in liquid helium. Redrawn from Schoepe and Rayfield, 1973.

The temperature and electric field dependence of the trapping time in an ionization chamber with a Frisch grid and electronically gated electron source (Fig. 4.4) by Schoepe and Rayfield (1973) allowed for the measurement of the size of electronic bubbles and the binding energy in liquid ^{4}He and ^{3}He ionized by ^{210}Po α-particles. By applying suitable electric fields with delays between the cathode and the gated grid, and between the anode and the Frisch grid, electron trapping times ranging from 0.4 to 100 s were measured in liquid He^4, He^3, and He^4+He^3 mixture.

4.3 Emission ionization chambers using organic liquids

The first observation of excess electron emission from a room temperature non-polar dielectric liquid was performed by Minday *et al.*

(Minday *et al.*, 1971) with a simple triode parallel-plate ionization chamber partly filled with liquid n-hexane in which a threshold-like behaviour of the IV-curve was observed. Electrons were injected by illuminating the cathode with a mercury lamp and drawn upward to the grid. Both the grid and cathode were immersed in the liquid. The grid was used to separate effects from charge carriers generated in the liquid from carriers emitted from the photo-cathode. Applying voltages between the cathode, the grid and the anode located in gas above the surface of the liquid, Minday *et al.* demonstrated that electrons could be extracted from liquid n-hexane at electric fields of > 50 V/cm strength. In contrast, positive ions could not be extracted with electric fields up to 500V/cm.

The emission of excess electrons from liquid iso-octane was discovered by Balakin *et al.* (1977) with a parallel-plate ionization chamber irradiated with 1 MeV electron beam from an accelerator (Fig. 2.11). Observations of the electron emission kinetics led Balakin *et al.* (Balakin *et al.*, 1977) to conclude a thermal origin for the electron penetration through the surface barrier as described in Section 2.3.2.

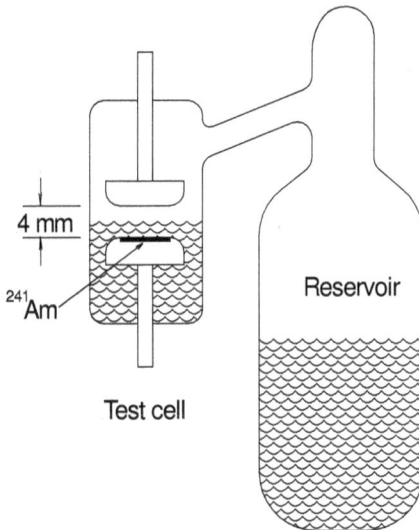

Figure 4.5 Schematic drawing of the glass ionization emission chamber filled with TMP. Redrawn from Anderson *et al.*, 1987.

In 1987, Anderson *et al.* observed electron emission from liquid argon and discovered electron emission from 2,2,4,4-*tetrametylpentane* (TMP) accompanied by gas gain in the equilibrium gas phase. The glass ionization chamber (as shown in Fig. 4.5) included two 15 mm diameter electrodes separated by a gap of 4 mm with the lower electrode coated with an ^{241}Am alpha-particle source. The ionization chamber was also tested with liquid argon for comparison.

For tests with liquid argon (LAr), a pulse height spectrum was taken for two cases. At first, the gap was filled with LAr and an extraction electric field 3.2 kV/cm was applied. A peak was measured with energy resolution 25% FWHM. For the second test, the LAr level was lowered to the middle of the gap. At the same voltage, the energy resolution was measured to be 19% FWHM. The improved energy resolution may be a result of the reduced number of electrons lost during the drift through the thinner layer of the liquid argon. No stable gas gain was observed in the argon vapour.

For tests with TMP, the chamber was connected by a thin glass tube to the glass reservoir containing liquid TPM to allow the level of the TMP in the chamber to be adjusted. After purification, the TMP was placed in the baked-out glass chamber and the chamber was sealed off after 80 torr argon gas was added to create a mixture with TMP vapour having 14 torr pressure at 20°C. The electron lifetime in the TMP was 7 μs. In this experiment, an effective gas gain was observed and is discussed in Section 5.2.

Chapter 5

Emission Detectors with Physical Amplification of Signals

Physical amplification of primary ionization signals can improve the performance of ionization detectors. Earlier attempts to achieve effective charge or light multiplication in noble liquids in the vicinity of thin wires were not very successful (Derenzo *et al.*, 1974; Miyajima *et al.*, 1976). In response, micro-strip boards (Akimov *et al.*, 1994; Policarpo *et al.*, 1995) and chemically etched needles (Kim *et al.*, 2004) were investigated but achieved only low amplification factors (of about 10–100) and poor energy resolutions. In rarefied phases (low density gas or vacuum), however, electron signals can be amplified relatively easily, which is a major advantage of emission detection technology.

Figure 5.1 shows the band diagram of a *diode emission detector*. Quasi-free electrons generated by radiation absorbed in the condensed phase lose their original energy by creating ionizations and excitations in the condensed medium (at relaxation length χ), drift to the interface surface, penetrate the surface potential barrier Δ and escape into the rarefied phase (gas or vacuum). In vacuum, electrons are accelerated by the electric field $F2$ before reaching the anode. The energy acquired from the electric field $E3$ can result in secondary ionizations on the anode and, thus, amplify the electron signal. In gas, drifting electrons can acquire energy $E2$ which can be high enough to excite the gas and generate electroluminescence or result in avalanche multiplication of the electrons. Both processes can be used for amplifying the electron signal. Additional $E2 + \varphi^a$ energy gain is possible if superconductive anodes connected to transition-edge sensors (TES) can be used in order to collect ionization electrons.

In this chapter, we will consider the most practical ways of physically amplifying the emission electron signal.

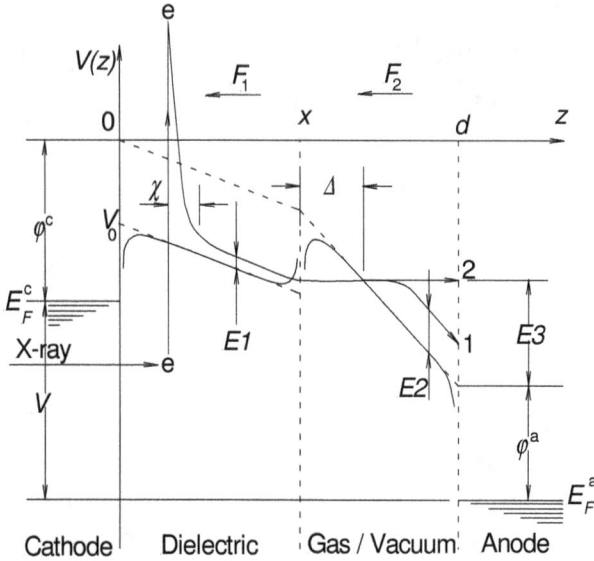

Figure 5.1 Band diagram of diode emission ionization chamber with non-polar dielectrics working medium of thickness x and between-electrode gap d biased with voltage V.

5.1 Emission detectors with acceleration of electrons in vacuum

The absence of diffusion in vacuum, high rate of transition of electrons, and possibility of easy transformation of electronic images make emission detectors with solid noble gas very attractive. A necessary condition of such detectors is a low ($< 10^{10}$ cm^{-3}) vapour density or low ($< 10^{-4}$ Pa) pressure. The advantage of electron collection in vacuum is the possibility of accelerating electrons and supplying them with sufficient energy for detection of single electrons.

The effect of *secondary electron emission* from metals has been employed in parallel plate vacuum ionization chambers for the detection of high-intensity beams of high-energy particles (see, for example, Hori, 2004). Since the electric field can not penetrate deeply inside metals, the escape depth of electrons from metals is limited with their thermalisation length, which is reasonably high only for ionization by accelerated particles. In p-type silicon with negative surface electron affinity (NEA), the thermalisation length of photo-electrons may reach 10 microns and such emitters can be used as thick photo-cathodes for the detection of relatively long-wave (infrared) photons (Martinelly and Fischer, 1974). However, such thin metals and NEA semiconductors are ineffective for the detection of penetrating radiation.

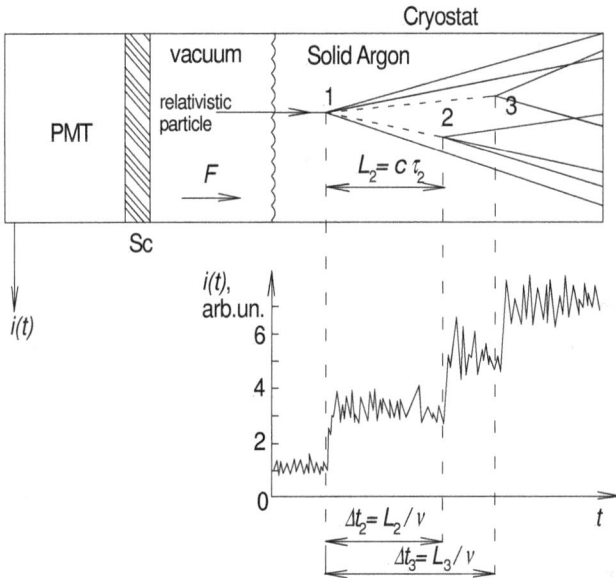

Figure 5.2 Concept of solid Ar-vacuum vertex detector for the detection of short-ranged neutral relativistic particles in an accelerator experiment: (PMT) photo-multiplier; (Sc) plastic scintillator; interaction point (1); point of generation of short-range neutral particles (2, 3). Redrawn from (Guschin, 1981).

Porous layers of micro-crystal *ionic crystals* such as MgO and CsI with grain sizes of up to 10 μm and densities consisting of only 0.7% of density of the bulk crystal can operate as effective emission detectors at sufficiently high electron extraction fields (> 100 kV/cm) for the multiplication of emitted electrons in pores. Such emission detectors have been proposed as fast, highly sensitive detectors of high energy particles in accelerator experiments (Lorikyan and Trofimchuk, 1977; Arvanov *et al.*, 1981).

As we have shown in Chapter 2, ionization electrons can be extracted from massive bulk non-polar dielectrics making them the most attractive media for emission detectors of single and low-ionizing particles. An attempt to construct a fast solid argon emission detector as a vertex detector for accelerated particles experiments was undertaken by E. Guschin, S. Somov, and B. Dolgoshein at the end of the 1970s (Guschin, 1981). The operating principle of the time-projection emission detector is illustrated in Fig. 5.2. Electrons extracted from tracks of ionizing relativistic particles were shown to be extracted from the solid argon, accelerated in the vacuum and detected by a plastic scintillator viewed by a fast PMT. Analyses of the waveform of signals acquired from the PMT allowed the measurement of the range of short-living neutral particles. For example, if the lifetime of a short-ranged neutral particle is 10^{-13} sec, it travels $L = 30$ µm in solid argon before decaying into charged particles. This feature in the distribution of ionization electrons transforms into a current pulse of $\Delta t = 3$ ns duration (for electron drift velocity $v = 10^6$ cm/s in solid argon) that can be measured by the PMT. Unfortunately, the performance of the instrument was poor because of the low mobility of ions in the solid argon leading to their fast polarisation in the intensive beam of high-energy particles and because of discharges in vacuum which paralysed PMT operation.

However, we believe that this approach may be more fruitful in the future for low rate applications. Another intriguing possibility that has not yet been realised is the development of an *emission microscope* using relatively easy transformations of electron images in vacuum by electronic optics similar to the technique used in electron microscopes (for information on electron microscopes, see (Kobayashi *et al.*, 1999). This kind of emission detector may be used for investigating ionizing particle track structure (for background rejection, for example) or for collecting data from a large area detector to relatively small readout devices such as an array of charge-coupling devices (CCD).

5.2 Emission detectors with gas gain

Effective electron avalanche multiplication (gas gain) may be used in emission detectors with saturated vapours of condensed mediums, ballast gas placed above condensed detection mediums or a mixture of gases.

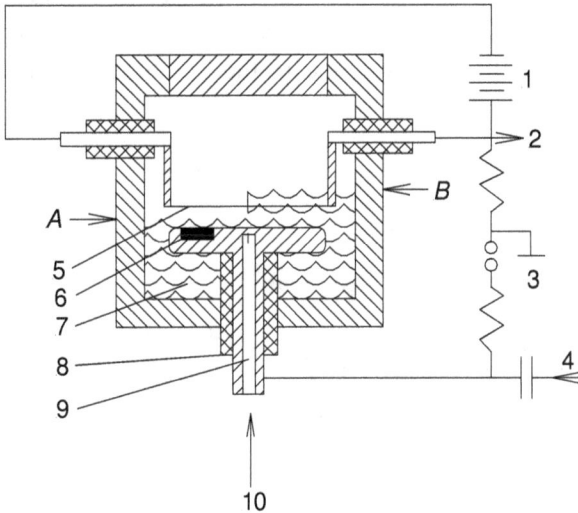

Figure 5.3 Emission wire proportional chamber with multi-wire anode operating above (A) or under (B) the free surface of liquid argon: battery heating wires with 0.3 A current (1); output anode signal (2); DC high-voltage (3); pulsed high voltage (4); anode composed of 100 micron wires (5); alpha particle source (6); liquid argon (7); HV feedthrough (8); collimator (9); input radiation (10). Redrawn from Dolgoshein *et al.*, 1973.

One of the first such emission detectors was a two-electrode liquid argon ionization chamber with a multi-wire anode (Dolgoshein *et al.*, 1973). The operating principle of this detector is illustrated in Fig. 5.3. As seen in this figure, the anode consisting of 50–200 μm diameter wires could be placed above (A) or under (B) the liquid surface. Unstable electron multiplication with a gas gain of less than 500 was observed in the first case. In liquid argon, the anode wires were heated by an electrical current of 0.1–1.0 (A) in order to boil the liquid to form a vapour jacket around the wires. In this case, electrons were emitted inside the vapour jacket and avalanched in the vicinity of the wires. The avalanche length was limited by the size of bubbles forming the gas jacket. A gas gain of 10^4 was achieved by increasing the dead time (10 ms against 0.1 ms in the vapour). The elevated dead time was probably associated with the localisation of positive ions inside bubbles and by the essential time needed for the collection of the ions. With a pulsed HV supply, the gain around heated wires in liquid argon was increased up to 10^6. However, the method required a huge amount of electrical power.

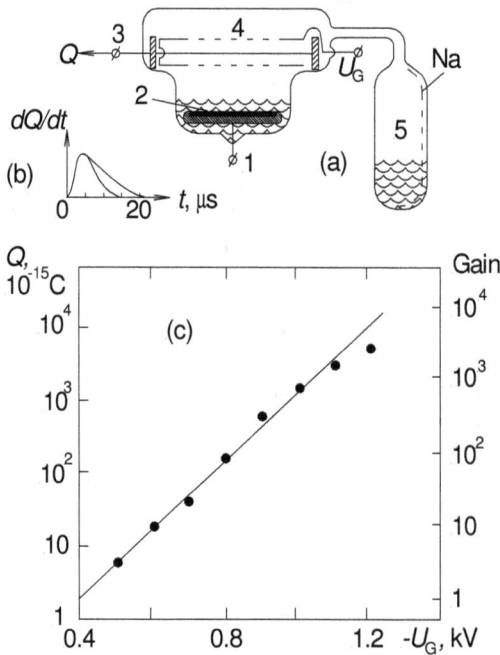

Figure 5.4 Emission wire proportional chamber with liquid iso-octane working medium operated at room temperature (a), typical waveforms (b), and gas gain curve (c): cathode (1); alpha-particle source (2); multi-wire anode (3); grid (4); storage reservoir with sodium getter (mirror) deposited onto inside surface (5). Redrawn from Bolozdynya *et al.*, 1978.

Another important example of an emission chamber utilising gas gain is a wire proportional chamber with liquid iso-octane (2, 2, 4 - trimethylpentane) operating at room temperature. Emitted electrons multiplied near anode wires in saturated iso-octane vapour at ~ 100 torr pressure (Bolozdynya *et al.*, 1978). The design of the sealed-off glass chamber used in this study is shown in Fig. 5.4a. The iso-octane was ionized with a 30 mm diameter alpha-source installed on the cathode immersed in the liquid. The anode was made from a planar 30 μm nichrome wire grid with a pitch of 3.5 mm. At a distance of 4 mm from both sides of the anode, two flat screening grids made of 200 μm diameter nickel wires with a pitch of 0.8 mm have been installed. The wire chamber formed by the anode and two screening grids was installed above the liquid. The iso-octane was purified with molecular sieves and a sodium thin film getter. In addition, a sodium getter mirror was deposited onto the walls of the storage reservoir attached to the detector. The brightness of the sodium mirror served as an indicator of the purity of the liquid. The thickness of the liquid inside the detector was controlled by removing or adding the liquid from the reservoir. During the six months of operation, no aging effect was observed; the 7 μs lifetime of quasi-free electrons in the liquid was constant. Stable proportional electron multiplication with a maximum gain of 30,000 was achieved. Figure 5.4b shows the gain as a function of applied voltage.

A gas gain of over 10^3 has been observed in a parallel-plate two-electrode emission chamber filled with 2,2,4,4-tetramethylpentane (TMP) operating at room temperature (Fig. 5.5). Figure 4.6 shows a diagram of the glass sealed-off detector used in this study. A ballast gas of 80 torr argon was added to the TMP vapour (14 torr at $20°C$) in order to increase the gas pressure above the liquid and improve the electron multiplication. The electron lifetime in the liquid was measured to be 7 μs and was not degraded by the addition of argon.

The electron multiplication in krypton vapour has been studied for the detection of radiation in a small emission detector. In this experiment, the 13 mm diameter central part of the anode was equipped with a grid consisting of 20 μm tungsten gold-plated wires with a pitch of 1 mm (Anisimov *et al.*, 1986). The detector operated with either pure krypton and krypton–methane mixtures. The thickness of the liquid was

1 mm in all cases. X-ray radiation with a maximum energy of 35 keV was used to irradiate the liquid. With pure krypton vapour, a gas gain of only 60 was observed above the liquid krypton. In contrast, with the mixtures of Kr+5mol% CH_4 and Kr+39mol% CH_4, gas gains of 200 and 500, respectively, were observed. Using this detector, it was shown that the threshold of the emission increases with increasing concentrations of methane due to the electrons being cooled by collisions with methane molecules.

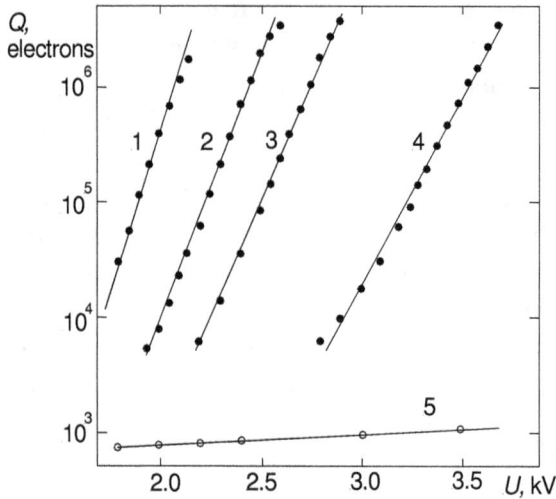

Figure 5.5 Number of electrons in TMP vapour in mixture with 80 torr (at 20°C) of argon measured by a parallel-plate emission ionization chamber filled with liquid TMP at 24°C (1), 30°C (2), 40°C (3), 50°C (4) as a function of voltage. Also shown for comparison is the ionization yield curve (5). Redrawn from Anderson *et al.*, 1987.

Gas gain in pure noble gases is unstable and limited to relatively low values because of secondary effects, such as the generation of UV photons and the photo-emission of electrons from electrodes developing into uncontrollable discharges. In an attempt to suppress the photon feedback, gas electron multipliers (GEMs) have been tested in emission detectors (Bondar *et al.*, 2004). The operating principle of an emission detector with GEM readout is shown in Fig. 3.16. The three-layer GEM of 28 mm by 28 mm sensitive area was made out of a 50 μm thick piece of Kapton with 70 and 55 μm hole diameters on the copper coating and insulating foil, respectively, at 140 μm pitch. The distance between the GEM1 and the chamber bottom is 5 mm. The distances between GEM1, GEM2, and GEM3 are 2 mm. The signals were recorded from the last layer (GEM3) in either the current or pulse-counting mode. In the latter case, a charge-sensitive amplifier was used with a 10 ns rise time, 8 μs decay time and sensitivity of 0.5V/pC. The detector operated as an emission device filled with liquid krypton and was irradiated with β-particles from a ^{90}Sr source.

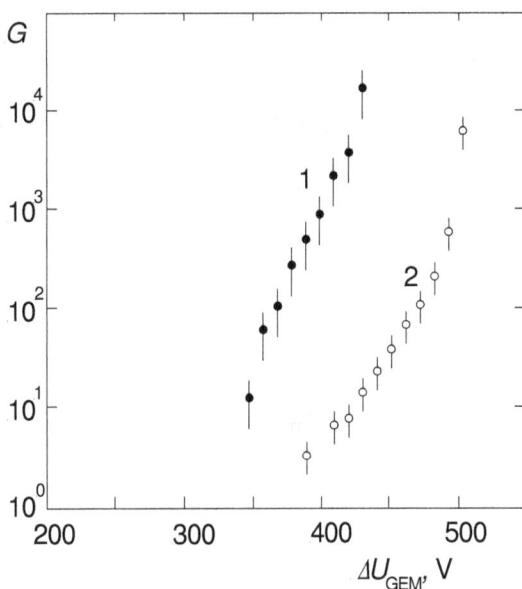

Figure 5.6 Gas gain as a function of voltage in a triple-GEM electrode system of liquid krypton emission detector as shown in Fig. 3.16 operating at 118 K and 0.84 atm vapour pressure (1) and at 123 K and 1.32 atm vapour pressure (2). Redrawn from Bondar *et al.*, 2004.

As seen in Fig. 5.6, the triple-GEM electrode system allowed gas gains of up to 10^4 in pure krypton vapour operating in two-phase mode. In a LXe emission detector, such a high gain can be achieved only with an admixture of CH_4 (Fig. 3.18). With pure krypton gas at room temperature gains of up to 10^5 have been achieved (Bondar *et al.*, 2004).

5.3 Emission detectors with light amplification

Emission detectors with electroluminescence of the gas phase for signal amplification are currently the most popular two-phase emission detectors. In general, data from the simplest two-electrode emission ionization chamber can be taken in two channels by measuring the collected charge (as described in Chapter 4) and/or measuring the photon emission when ionization electrons are collected in high enough applied electric field to generate electroluminescence of the gas phase.

Using the modified emission detector shown in Fig. 5.3 (the anode made as a flat grid of parallel wires of 50 μm diameter with 600 μm pitch and a PMT installed above the window), Dolgoshein, Lebedenko and Rodionov (1970) triggered the emission detector with scintillation initiated in liquid argon by alpha-particles and observed the electroluminescence flash when extracted ionization electrons drifted through the gas phase. They pointed out that the delay time between the scintillation flash and the electroluminescence signal can be used to determine the depth position of the alpha-source. This was the first electroluminescence emission detector operating as a time-projection chamber. Soon afterward, Abramov *et al.* (1975) demonstrated that it was possible to identify radiation of different origins (as shown in Fig. 5.7) by analysing waveforms of electroluminescence signals.

Lansiart *et al.* (1976) used a similar approach to detect γ-rays in liquid xenon (Fig. 5.8) and suggested that emission detectors could be used for imaging γ-ray fields in nuclear medicine because the electroluminescence flash provides a much better light output than the scintillations in crystal scintillators normally used in gamma cameras for medical imaging (see Chapter 6 for details).

A miniature electroluminescence emission detector made from a piece of glass tube with a wavelength shifter deposited inside (as shown in Fig. 5.9) has been used to study the threshold field effects associated with the hot electron emission from solid argon (Bolozdynya, Miroshnichenko *et al.*, 1977), liquid and solid krypton, methane (with neon as a ballast gas) and their mixtures as well as for comparing scintillation and electroluminescent properties of different condensed noble gases (Bolozdynya, 1986).

The advantage of the *two-mode readout* from LXe emission detectors for dark matter experiments has been investigated with a three-electrode parallel-plate emission detector (Aprile *et al.*, 2004) equipped with a metal-body quartz-window Hamamatsu R6041 PMT (20% quantum efficiency at 178 nm) as shown in Fig. 5.10a. A 6 cm

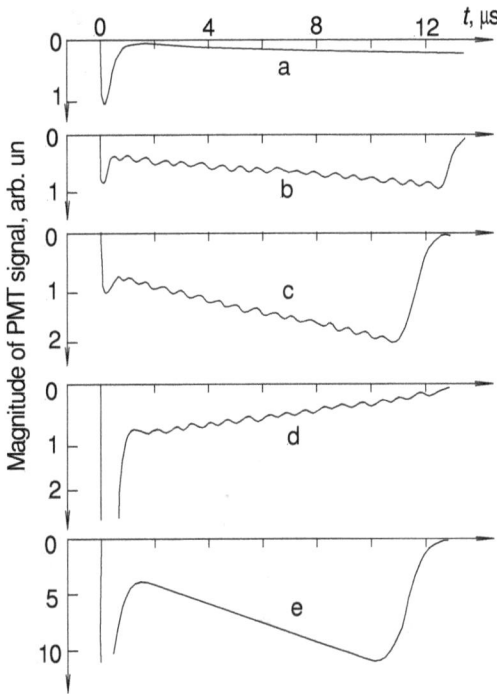

Figure 5.7 Waveforms of luminescence signals acquired from two-electrode parallel-plate emission ionization chamber filled with ~ 1 mm layer of solid xenon and gaseous neon at ~ 100 K and irradiated with different radiations: scintillations from alpha-source deposited onto the cathode (a); scintillation signal continuing with electroluminescence signal generated by extracted electrons in neon at 3.3 and 4.7 kV/cm electric field in the solid xenon (b, c); and signals from cosmic muons crossing the detector when extraction electric field is applied in inverse (d) and normal (e) directions. Redrawn from Abramov *et al.*, 1975.

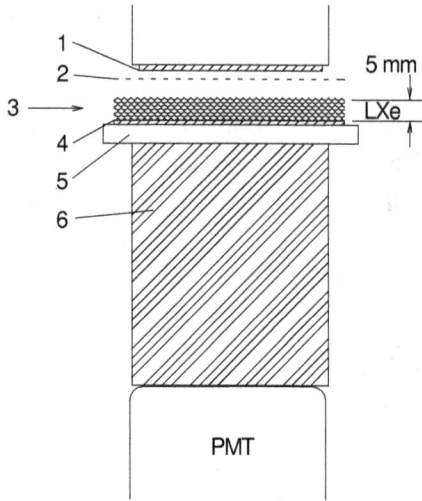

Figure 5.8 Electroluminescent emission LXe detector with photo-multiplier readout: cathode (1); grid (2); 5 mm diameter γ-ray beam (3); transparent anode (4); glass window (5); Plexiglas light guide (6). (PMT) photo-multiplier 53 AV installed with optical grease. Redrawn from Lansiart *et al.*, 1976.

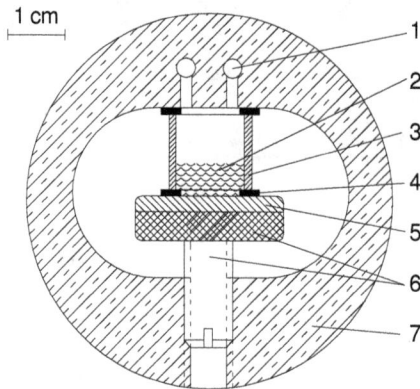

Figure 5.9 Glass-tube electroluminescent emission detector with photo-multiplier readout (Bolozdynya *et al.*, 1977): gas inlet and outlet (1); condensed phase (2); 15 mm diameter glass tube coated inside with p-terphenyl wavelength shifter (3); Teflon foil gasket (4); cathode coated with alpha-source (5); sealing screw with insulating pad (6); thermal screen (7). A photo-multiplier with 52 mm input window (not shown) was installed behind the glass cell.

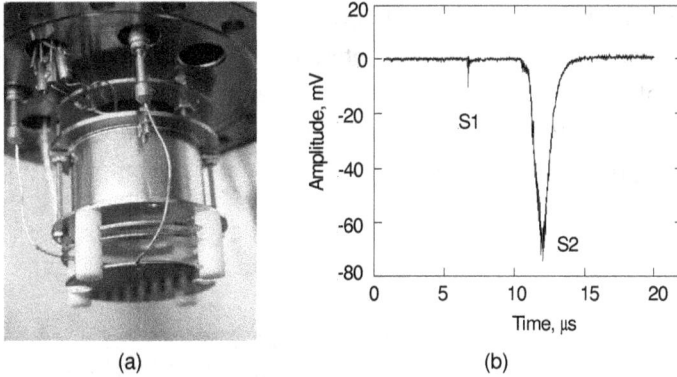

Figure 5.10 Electroluminescence emission detector with a photo-multiplier placed in the gas phase (a) and typical waveform (b) measured from the PMT. Courtesy of E. Aprile (a) and redrawn from Aprile *et al.*, 2004 (b).

diameter stainless steel disk with a ^{207}Bi β-source deposited at the centre served as the cathode. The grid structure was mounted 10 mm above the cathode and consisted of 0.1 mm wires at 1 mm pitch. A typical waveform measured from the PMT is presented in Fig. 5.10b. The first signal S1 is associated with the scintillation emitted promptly after the particle interacts with the liquid xenon and the second signal S2 is from the electroluminescence flash in the Xe vapour when the cloud of ionization electrons drift through the gas. The energy spectra of the electroluminescence signals S2 was compared with the ionization signal from the detector functioning as a LXe ionization chamber. There is a peak at about 80 keV in S2 spectrum while the low-energy section of the charge spectrum is dominated by noise. This clearly demonstrates the advantage of emission detectors against ionization chambers by being able to amplify the acquired signals. The detector served as a prototype for the XENON-10 emission detector used for cold dark matter search as described in Chapter 7.

Modern emission detectors contain many kilograms of LXe or LAr and use arrays of PMTs to collect scintillation and electroluminescence photons from the bulk sensitive volume. For example, the ZEPLIN-III detector (Fig. 5.11) achieved a low threshold for the detection of the primary scintillation by placing a PMT array in the liquid. Placing the PMTs in the liquid improves the light collection by removing two optical interfaces, both with large refractive index mismatches, and takes advantage of the large amount of internal reflection. A low threshold for the electroluminescence from the gas phase is achieved by using a high electric field in the gas region and by using refraction at the liquid surface to produce a 'focussing' effect on the light onto the immersed PMT array.

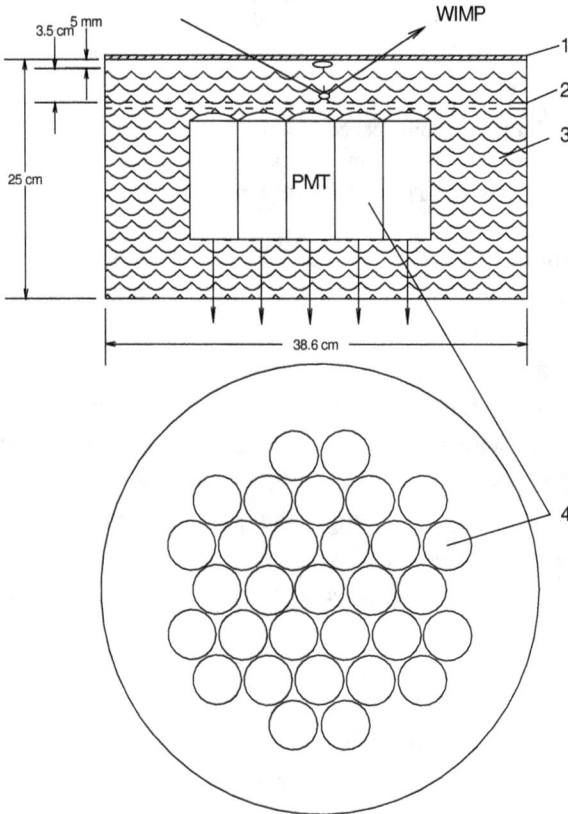

Figure 5.11 Schematic drawing of the ZEPLIN-III electroluminescence emission detector: highly reflective anode (1); cathode and grid screening PMTs (2); liquid xenon (3); array of 31 hexagonally packed 2" diameter photo-multipliers (4); WIMP – weakly interacting massive particle generating a scintillation flash followed by an electroluminescence flash. Redrawn from Akimov *et al.*, 2007.

5.4 On the possibility of amplification using transition-edge sensors

The detection threshold is normally limited by *electronic noise* which, for all aforementioned emission detectors, exceeds 10 keV. According to Radeka (1988), only thermal fluctuations in the dielectrics (for example, in printed circuit boards) generates noise

$$ENC_D \cong \left(2.4 k_B T \cdot D \cdot C_D\right)^{1/2} \qquad (5.1)$$

where k_B is the Boltzmann constant, T is the temperature, the dissipation factor for the best dielectrics is $D \cong (5\text{-}10)\ 10^{-5}$, which results in $ENC_d \cong$ 50–100 rms electrons at $C_D = 10$ pF. This noise can not be eliminated for devices working at room temperature. The only way to reduce the electronics noise and associated detection threshold is to develop emission detectors working at cryogenic temperatures.

Following the above considerations (Section 3.6), cryogenic emission detectors coupled to transition edge sensors may have a detection threshold of < 10 eV. The detector architecture could be similar to that used in the *CDMS-II* detector and the proposed *SuperCDMS* experiment (Schnee *et al.*, 2005), utilising arrays of cryogenic germanium and silicon detectors. The detector may consist of a solid xenon time-projection chamber where the ionization electrons are collected from the bulk solid xenon onto superconductive aluminium anodes coupled to the transition edge sensors (Bolozdynya, 2005). Since the drift length of quasi-free electrons in solid xenon already have been observed to be more than 1 m, the mass of the working medium can be up to a few tonnes. The detector should be surrounded by several heat screens connected to a dilution refrigerator cooling the detector to below 100 mK.

Thus, further enhancements of the emission detector technology may be associated with the exploration of solid working media such as solid xenon operating at cryogenic temperatures.

Chapter 6

Imaging Emission Detectors

Due to the inherently large volume of the working media, emission detectors operating with non-polar dielectrics are capable of determining three-dimensional distributions of charge carriers generated by radiation, i.e. for imaging ionization traces left by high-energy radiation fields and by low-ionizing single particles. There are two basic approaches to reconstructing the spatial distribution of ionization in the bulk detection medium and they may be classified as analogue imaging and digital imaging. *Analogue imaging* means that the charge distribution is visualised and recorded with standard optical imaging devices, such as photo, video or CCD-cameras. *Digital imaging* means the creation of digital images in result of recording, storage and computer processing electrical signals in order to reconstruct images in 'filmless' manner.

6.1 Analogue imaging cameras

In analogue imaging detectors traditionally called 'cameras', original ionization electrons are used to generate light with enough intensity to be captured with standard optical image recording equipment.

The first imaging emission detector was developed during R&D for an imaging *streamer chamber* based on liquid noble gases (Dolgoshein, Lebedenko and Rodionov, 1967). Emission detector features were first considered in detail by Rodionov (Rodionov, 1969–1987) and the results of first experiments were published by Dolgoshein, Lebedenko and Rodionov (1970).

6.1.1 Emission spark chamber

Since, at elevated voltages, it is relatively easy to develop sparks in pure noble gases (Dolgoshein, Lebedenko and Rodionov, 1967), single electrons extracted from noble liquids or solids can ignite sparks that are visible by naked eye. That discovery opened the possibility to produce visible tracks of individual particles passed through condensed detector media.

Sparks in vicinity the anode wires have been used to image α-particle source mounted on the cathode in the liquid of an emission detector similar to one shown in Fig. 5.2 (Dolgoshein, Lebedenko and Rodionov, 1970). The anode was made as a flat grid of parallel wires of 50 μm diameter with 600 μm pitch. The gap between the anode and the cathode was 1.4 cm, the thickness of the liquid argon was 0.4 cm. Neon was added and the gas phase above liquid argon was composed of 50% argon vapour and 50% neon. Ionization electrons created in liquid argon by α-particles emitted from the disk source of 3.5 cm diameter were extracted from the liquid by a 3 kV/cm electric field. The sparks were generated by

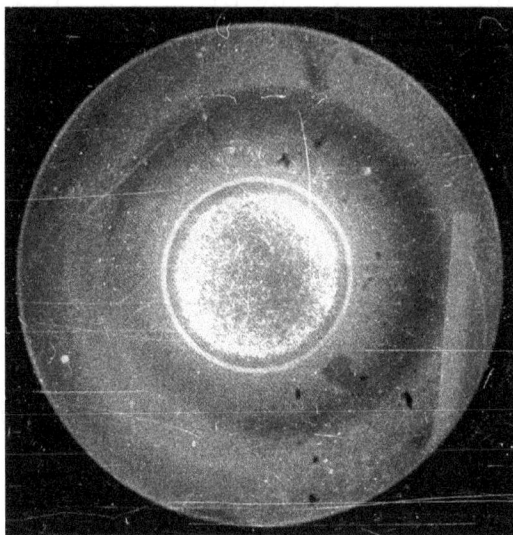

Figure 6.1 Image of alpha-source in the LAr emission spark chamber (Dolgoshein, Lebedenko and Rodionov, 1970). Courtesy of B.U. Rodionov.

a pulsed electric field of 40 kV magnitude and 100 ns pulse duration. The spark discharges were photographed with a camera through the window located above the wire anode.

The image of the disc alpha-source composed by superposition of images of many sparks is shown in Fig. 6.1. One can see that the distribution of the radioactive material over the source area is not uniform that is specific for radioactive sources made by evaporating a radioactive solute. The bright halo around the active region is built by a portion of the spark light reflected from the rounded well in the cathode where the alpha-source was mounted.

The emission spark chamber did not find practical applications; however, this development clearly demonstrated that emission detectors can operate as imaging radiation instrumentation.

6.1.2 Emission streamer chambers

The next modification of the spark emission chamber was an *emission streamer chamber* tested by a beam of relativistic particles at the ITEP synchrotron in 1977–1979 by a group lead by Rodionov (Bolozdynya, Egorov *et al.*, 1977, 1980). In this detector, a 0.5 cm layer of solid krypton was used as a working medium and a 1 cm gas gap filled with neon at normal pressure as an amplification medium (Fig. 6.2). The anode had a 12.5 cm diameter gridded central part allowing the gas gap to be photographed through the glass window installed above the anode. The window was used as a high-voltage insulator supporting the anode. The bottom of the detector vessel served as a cathode, the form of which provided a uniform electric field between the gridded part of the anode and the cathode (the extra-volume was filled with massive Teflon displacer). A positive DC voltage was applied to the anode to provide the emission of electrons at 1.5 kV/cm electric field strength. Triggered by a telescope of external scintillation counters, a high-voltage pulse of 60 ns duration was applied to generate the streamer discharges along high-energy particle tracks extracted from the solid krypton.

The streamer discharge is developing when the magnitude of the electric field exceeds 2 kV/cm in the gas phase. Surprisingly, the strength

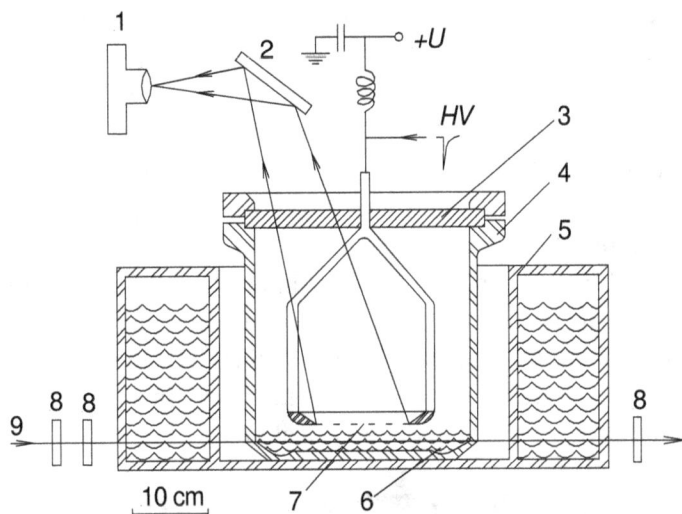

Figure 6.2 Emission streamer chamber: photo camera (1); mirror (2); window (3); vessel (4); liquid nitrogen thermostat (5); condensed krypton (6); gridded HV electrode (7); scintillation counters (8); track of relativistic particle (9). Redrawn from Bolozdynya *et al.*, 1980.

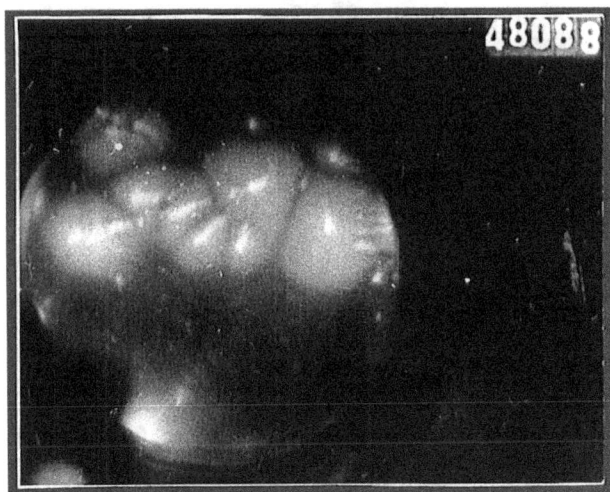

Figure 6.3 Image of relativistic particles interactions taken with the first emission streamer chamber. The diameter of the field of view is 12 cm; luminous areas around streamers are due to the leaking of the discharge plasma over the surface of the solid krypton.

of the 'visualising' pulsed electric field was found to be about the same as that used to operate the streamer chamber at room temperature; I. Sidorov came to the same conclusion testing helium cryogenic streamer chamber operating at 4 K a few years earlier (Gorodkov *et al.*, 1974; Sidorov, 1975). The effect may be associated with increasing concentration of noble atom dimers at cryogenic temperatures that lead to reduced effective ionization potential of the gas.

Tracks of 3 GeV/c pions and their interaction (Fig. 6.4) formed by chains of streamers have been photographed with single photo-cameras or with two photo-cameras for stereo imaging. The most photographed images were straight tracks but sometimes images of particles interactions have been captured. A typical image of a straight 'emission'

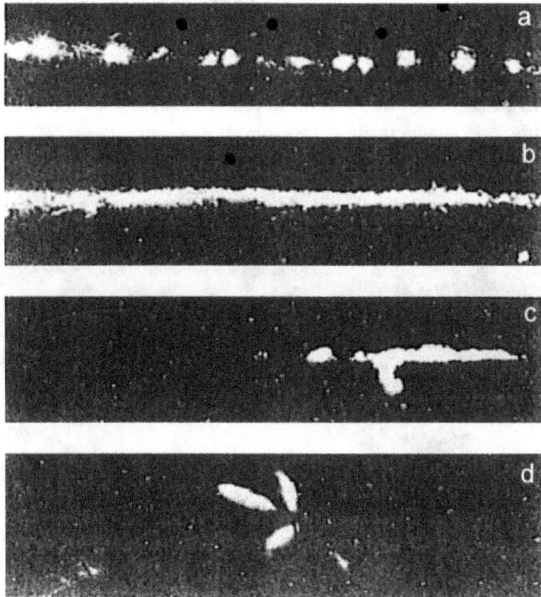

Figure 6.4 Tracks of ionization particles acquired from the emission streamer chamber: track of relativistic particle in the gas phase (a); tracks of relativistic particle (b), delta-electron (c) and products of interaction of relativistic particle with the bottom of the detector (d) in solid krypton. The diameter of the field of view is 12 cm, thickness of solid krypton is 5 mm, temperature 78 K, extraction field 1.5 kV/cm. Picture taken with 35 mm photo camera Zenith using objective Jupiter-9 (83 mm focus) from distance of 1.6 m (Bolozdynya *et al.*, 1977)

track is shown in Fig. 6.4b. A *track* of relativistic particles passing through the gas phase is shown in Fig. 6.4a. The image of the emitted track looks as a nearly continuous line without separation into individual streamers. This is a result of very dense electron distribution along the emitted track. The tracks obtained in emission mode consisted of 2 mm long by ~ 0.5 mm diameter streamers packed with a density of about one streamer per 1 mm of the track length.

Observation of dense emission tracks guided the authors to the idea that such kind of detectors can be useful for the detection of hypothetical abnormally low-ionizing particles (Bolozdynya *et al.*, 1980). Rare events, like that shown in Fig. 6.4a, might be considered as possible evidence of the existence of abnormal particles producing ionization along their tracks with density in several orders of magnitude fewer than that of normal relativistic particles. However, it was recognised that the majority of observed 'abnormal' tracks are generated due to effect of memory of the detector. Sometimes, sparks generated close to the surface of solid krypton charged it for time comparable to the period of time between two consequent triggers. Then, the 'abnormal' track appeared as a 'phantom' image of the track visualised due to the previous trigger. In Fig. 6.3, it is clearly seen that plasma spots from neighbouring tracks interact on the surface of solid krypton.

Later, a large emission chamber was built to detect several π^0-particles generated during *anti-proton annihilation* in heavy nuclei (Barabash and Bolozdynya, 1993). The camera was designed with 0.5 m diameter by 20 cm deep condensed krypton working medium and 1.5 m diameter cryogenic streamer chamber placed above the condensed krypton for visualisation of particle tracks as shown in Fig. 6.5. The streamer chamber consisted of two sections located on both sides of the high-voltage electrode with gridded central part and supported by four large ceramic insulators. The high-voltage pulses of up to 250 kV magnitudes were generated by an Arkadiev–Marx generator (Egorov and Sidorov, 2000) with a shock capacitance of 2200 pF matched to the capacitance of the streamer chamber. The high-voltage pulse of 20 ns duration was formed with two discharge switches: two-electrode self-ignited switch 6 and three-electrode switch 12, triggered through the high-voltage cable delay line from the first cascade of the generator.

Emission Detectors

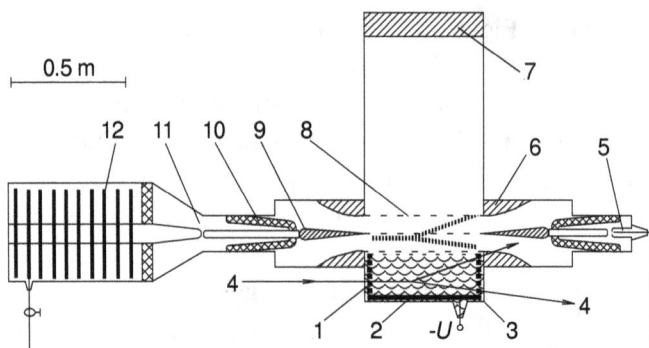

Figure 6.5 Streamer chamber 'Nadezhda': drift rings immersed in LKr (1); HV cathode (2); cryostat (3); particle tracks (4); triggered discharge switch (5); shaped ground electrodes (6); window (7); gridded electrodes of the streamer chamber (8); shaped HV electrode of the streamer chamber (9); ceramic insulator (10); passive discharge switch (11); Arkadiev-Marx generator (12); (U) drift voltage applied to the cathode; the streamer image of the tracks 4 is shown with vertical dashed line pattern between gridded electrodes 8.

The detector was constructed at ITEP (Fig. 6.6, top) in early 1980 and tested in parts: the whole-metal streamer chamber demonstrated a capability to detect tracks imitated by nitrogen laser (Fig. 6.6, bottom) and the 0.5 m diameter emission section of the detector was independently used as an ionization chamber for the measurement of krypton radioactivity and an electron emission purity monitor for large amounts of liquid krypton (Anisimov *et al.*, 1989a,b).

6.2 Digital imaging detectors

Since emission detectors produce electronic signals, digital signal processing can be performed. In digital electronic signal processing, charge-time distributions are digitized and image processing is performed with general-purpose computers using special algorithms to reduce noise and signal distortion that appear during the data taking. Depending on the technology of the physical signal amplification one can define a few approaches.

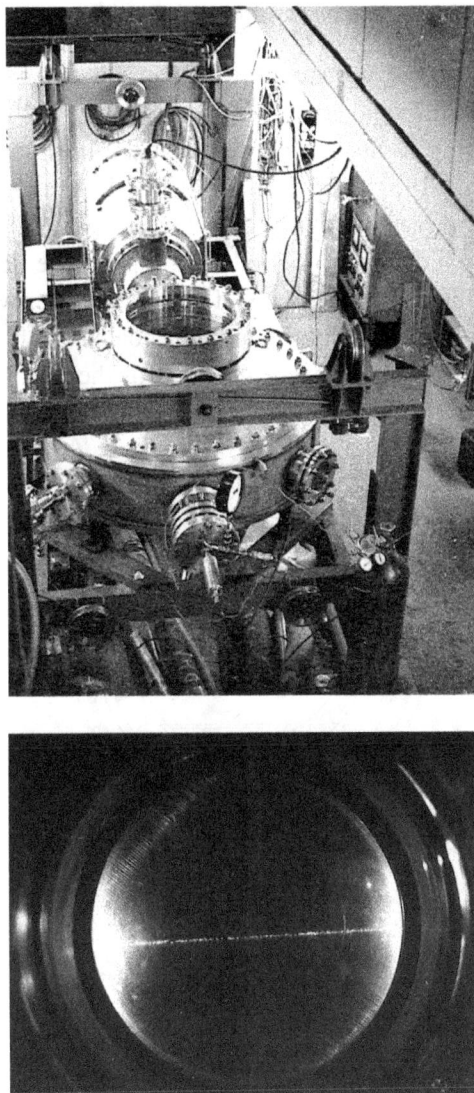

Figure 6.6 Emission streamer chamber 'Nadezhda' at the big experimental hall of the ITEP accelerator experimental hall (top). An image of UV laser beam imitating relativistic particle track in neon doped with low-ionizing alpha-naphtylamine vapour at room temperature and 1.2 bar pressure. The diameter of the field of view is 50 cm, the depth of view is 15 cm.

6.2.1 *Electroluminescence emission camera*

An *electroluminescence emission camera* with an array of photo-multipliers (Figs. 6.7 and 6.8) was developed for two-dimensional gamma ray imaging in nuclear medicine at the beginning of the 1980s (Bolozdynya *et al.*, 1981; Egorov *et al.*, 1983; Bolozdynya *et al.*, 1985). The stainless steel vessel of the detector enclosed a 30 cm diameter anode with 24.5 cm diameter gridded central part. The flat grid consisted of stretched, parallel nichrome wires of 50 μm diameter with 1 mm pitch. Nineteen 7 cm diameter glass windows coated with 0.5 mg/cm² p-terphenyl wavelength shifter were installed in the lid of the vessel in a

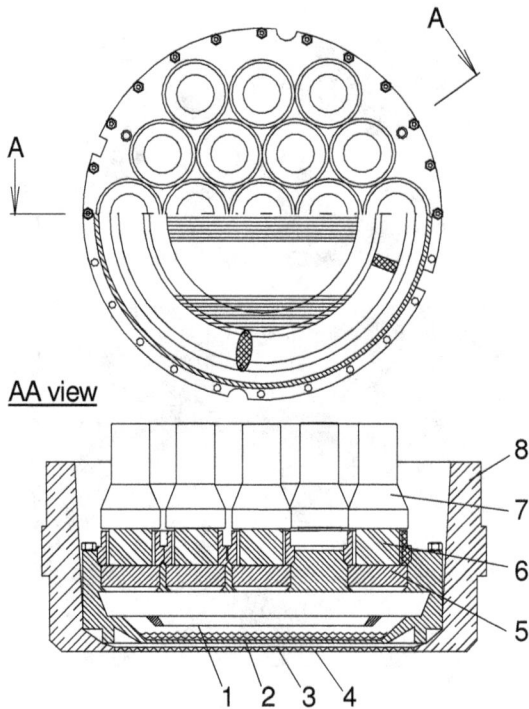

Figure 6.7 Electroluminescent gamma camera (Egorov *et al.*, 1983) with 19 photo-multipliers installed outside the detector vessel: gridded anode (1); cathode (2); vacuum insulation (3); thermal insulation (4); glass window coated with p-terphenyl wavelength shifter (5); acrylic light-guide (6); 3" photo-multiplier (7); part of the lead shielding (8).

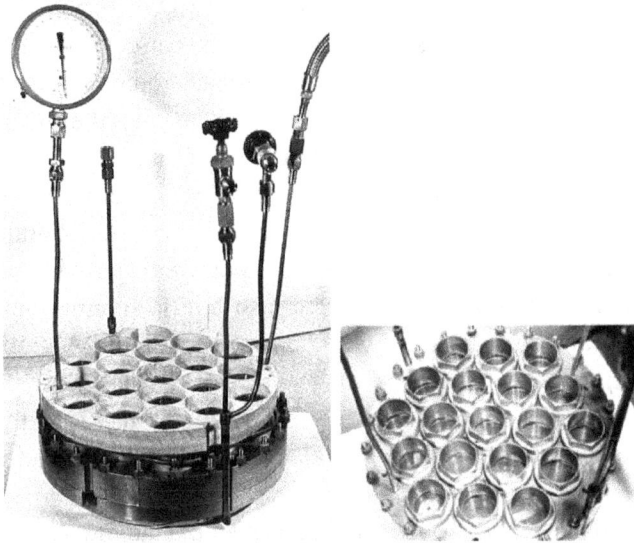

Figure 6.8 Electroluminescent gamma camera (Egorov *et al.*, 1983): a side view of the vessel with partly installed foam thermoinsulation (left) and a top view of 19 glass window array without thermoinsulation (right).

hexagonal pattern. Every window was viewed with individual FEU-110 glass photo-multipliers installed with acrylic light guides. Xenon or krypton was purified by passing through a hot (900 K) calcium absorber. The gas was purified with chromo-silicate oxygen adsorbent similar to the *Oxisorb* (a registered trademark of Messer Group GmbH for oxygen absorber based on silica gel activated with chromium salts). The gas was stored in a stainless steel tank (80 cm diameter by 150 cm tall), of which the inside surface was sputtered with a titanium getter generated by the head of a titanium sublimation pump installed on the top flange of the tank. The detector was cooled by nitrogen vapour circulating in the jacket surrounding the vessel. The gas was condensing in a layer up to 1 cm thick at the thin bottom of the vessel that served as a cathode. The detector was tested with krypton and xenon working media.

Coordinates of electroluminescence flashes were determined as position of centre-of-gravity of signals acquired from a PMT array operating in *Anger camera* mode (for details see Kalashnikov, 1985). A

Tektronix 603 storage oscilloscope was used to collect all data and to display images of an alpha-source placed on the cathode as well as a lead multi-hole collimator illuminated with gamma radiation from ^{241}Am or ^{57}Co gamma sources placed outside the detector; the images were photographed with a photo camera as shown in Figs. 6.9 and 6.10. Internal position resolution was measured to be 2.5 mm FWHM for the α-source installed on the cathode in solid krypton and 3.5 mm FWHM with the detector filled with liquid xenon and irradiated with a 59.6 keV (^{241}Am) gamma source. A pulse height distribution summarised over the PMT array analogue signals was analysed with an AI-256-6 pulse analyser; the energy resolution of 15% FWHM was measured with a ^{57}Co γ-source at 2 kV/cm electric field strength in 4 mm thick solid xenon and 16% FWHM at 4 kV/cm electric field in 1.5 mm thick sample of liquid xenon.

With this detector, electroluminescence of the liquid xenon was observed at uniform electric field of > 10 kV/cm field strength. Note that operations with krypton were affected by a high count rate (~ 10^5 s^{-1}) associated with radioactive decays of Kr^{85} radionuclide present in natural mixtures of krypton isotopes at the level of ~ 10^{-12} relative concentration (Anisimov *et al.*, 1989b). Operations with solid working media were

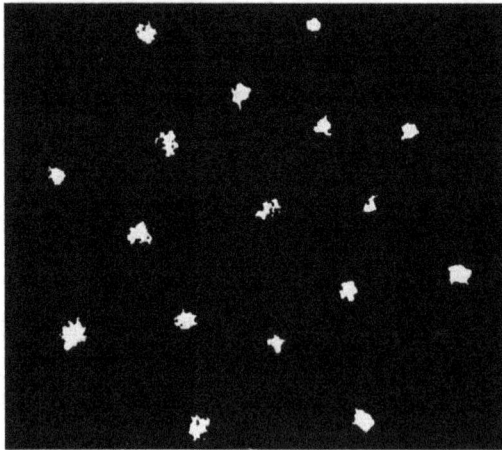

Figure 6.9 Image of lead mask over 22 cm field of view in gamma rays (Egorov *et al.*, 1983).

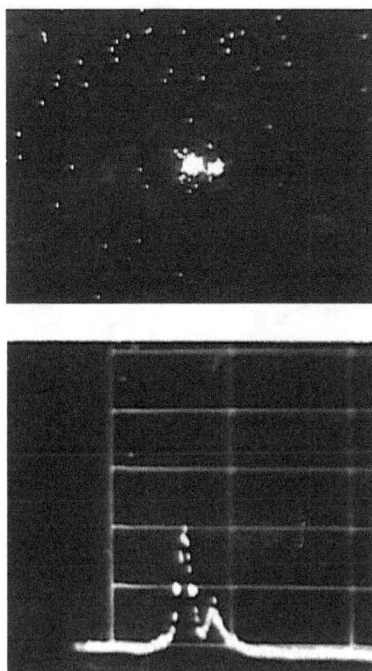

Figure 6.10 Image of lead mask with two holes of 1 mm diameter at 8 mm distance in 59.6 keV (^{241}Am) γ-rays (top) and distribution of count rate across the image; gas pressure 0.18 MPa, electric field strength 7 kV/cm (Bolozdynya, Egorov, Kalashnikov *et al.*, 1985).

possible only for limited time (~ 10 min.) because of polarisation of the working media due to low mobility of positive ions and holes. Depolarisation was achieved by application of the reverse electric field of the same strength for about the same time as operation in the normal mode.

Figure 6.11 shows the energy resolution of the emission electroluminescence camera detecting 59.5 keV (^{241}Am) and 122 keV (^{57}Co) γ-rays in solid and liquid xenon as a function of electric field strength in the condensed phase. One can see that the emission detector requires high electric field strength (4 kV/cm in liquid xenon and 2 kV/cm in solid xenon) to achieve the best energy resolution of 16 and 15% FWHM in liquid and solid xenon, respectively, irradiated with a ^{57}Co γ-source. The energy resolution is mostly limited by fluctuations of

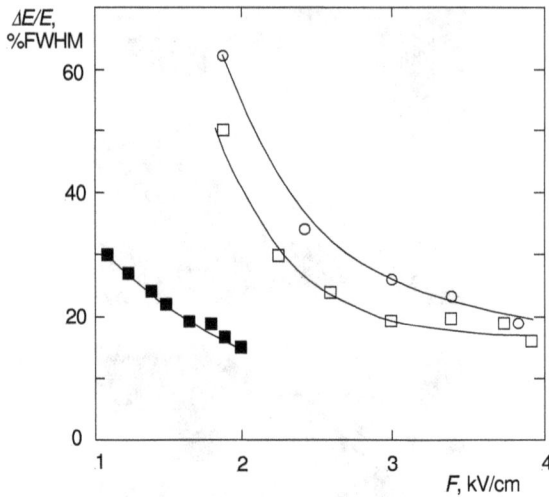

Figure 6.11 Energy resolution (% FWHM) of emission electroluminescence cameras as a function of electric field strength in liquid xenon (open signs; 167 K temperature, 1.5 mm thickness) and solid xenon (closed signs; 156 K temperature, 4 mm thickness) irradiated with 122 keV (squares) and 59.6 keV (circles) collimated to the centre of the field of view of gamma rays. Redrawn from Bolozdynya, Egorov *et al.*, 1982.

the electron yield from tracks of photo-electrons in condensed xenon (Bolozdynya, Egorov *et al.*, 1982).

6.2.2 *Emission time-projection chambers*

Emission detectors triggered by primary scintillation may operate in the time projection chamber mode for precise measurements of the depth coordinate of the working media or in the direction of the extracting electric field. One of the first of these detectors was developed by E. Guschin *et al.* as described in Section 5.1. I.M. Obodovsky with colleagues constructed the first electroluminescence time-projection liquid Xe emission drift chamber used in studies of the angular distributions of positron annihilation gamma-quanta (Guschin *et al.*, 1982). With a drift path of 2 cm, the detector demonstrated spatial resolutions of 0.5 mm at 30 keV and 1.0 mm at 511 keV in the direction of the electron drift. The resolution was mostly limited by the size of the diffused electron clouds collected in the gas. Similar small time-projection chambers detecting the both scintillations and electroluminescent signals with a single photo-multiplier have been used for discovery of hot electron emission from solid argon (Bolozdynya *et al.*, 1978), from liquid and solid krypton and from liquid and solid methane (Bolozdynya, 1986, 1999).

Recently, experimental collaborations searching for cold dark matter with emission detectors have constructed time-projection chambers as R&D devices with some of them achieving physical results (see Chapter 8). In this section, we will discuss the most typical examples of such detectors.

Single channel LXe TPC

In course of development of a detector for cold dark matter by the XENON collaboration, a small detector for initial R&D was constructed, as shown in Fig. 5.9 (Aprile, Giboni *et al.*, 2004). The cathode is a 6 cm diameter stainless steel disk with a ^{207}Bi β-source deposited in the centre. Two stainless steel etched meshes (0.1 mm bar width and thickness, 1 mm spacing) are mounted 10 mm above the cathode with 5 mm gap between them using Teflon insulators. The photo-multiplier is a Hamamtsu R6041 with 52 mm diameter quartz window and 4 cm long metal envelope with typical quantum efficiency < 8% at 178 nm. The detector was cooled to −95°C by immersion into an alcohol bath mixed with liquid nitrogen. The detector was used to optimise that data acquisition process and to test different photo sensors: photo-multipliers as described above, avalanche photo-diodes (Ni *et al.*, 2006), and silicon photo-multipliers (Aprile *et al.*, 2006). In result, photo-multipliers were selected as the basic technique for scintillation and electroluminescence photon detection in the XENON-10 and XENON-100 detectors.

Two-channel LXe TPC

A detector similar to the one described above was constructed by Shutt *et al.* (2006) but with an additional PMT installed below the gridded cathode that allowed improvement in the light collection efficiency. The active volume of this detector was Ø3.7 cm x 0.96 cm viewed with two Hamamatsu 9288 photo-multipliers. The light-collection efficiency of this detector was estimated at the level of ~ 50% resulting in detection of ~ 1 photo-electron per 1 keV deposited energy generated by nuclear recoils stopped in LXe. The trigger threshold was defined to ~ 5 electrons in the electroluminescence signal. The detector cryostat was a 'cold finger' type supporting the detector temperature with ±0.2 K accuracy. The liquid level between grids was measured with three open parallel-plate capacitors that allowed alignment at the level ~ 0.1 mm. The sensitivity of the capacitor level meter was high enough in order to detect ripples on the liquid surface generated by LXe droplets.

The detector was used to investigate the ionization and scintillation yield of LXe activated with low energy electrons and nuclear recoils (Shutt *et al.*, 2007), for testing p-terphenyl wavelength shifter (Bolozdynya, Bradley *et al.*, 2008), and charge multiplication in pure xenon (Dahl, 2009). It was shown that p-terphenyl wavelength shifter dissolving in LXe limits electron drift length below 0.5 cm at 1 kV/cm electric field strength.

In 2005, Suzuki *et al.* reported construction of two-channel emission time-projection chamber *XMASSII* with two photomultipliers operating in the coincidence mode and viewing 300 cm^3 LXe active volume. The detector has improved light-collection efficiency due to focussing Teflon reflectors installed as shown in Fig. 6.12. The anode and the cathode diameters and distance between them are 84, 46, and 85 mm respectively. The anode structure consists of two grids placed 5 and 15 mm apart as shown in the figure. The grids are made of 30 μm diameter wires with 2 mm pitch. Grounded cathode meshes with 90% optical transmittance are evaporated on the inner surface of the MgF$_2$ windows. Two UV sensitive PMTs (9426B by Electron Tubes, 32% QE at 174 nm and room temperature) are mounted outside the windows. A 40 liter liquid nitrogen bath was installed above the detector and used to cool the detector via a cold finger system adjusting the temperature to 180 K with heaters. The detector has been evacuated to 10^{-5} Pa and baked out for about three weeks at 120°C xenon gas was purified by passing through 3.3 L of Messer Grieshein Oxisorb. The electron lifetime in LXe was estimated to be > 1 ms and was maintained for three months of continuous operation without any additional purification. This result clearly demonstrated that massive (relative to the total mass of the working medium) Teflon reflectors can be used in sealed emission detectors.

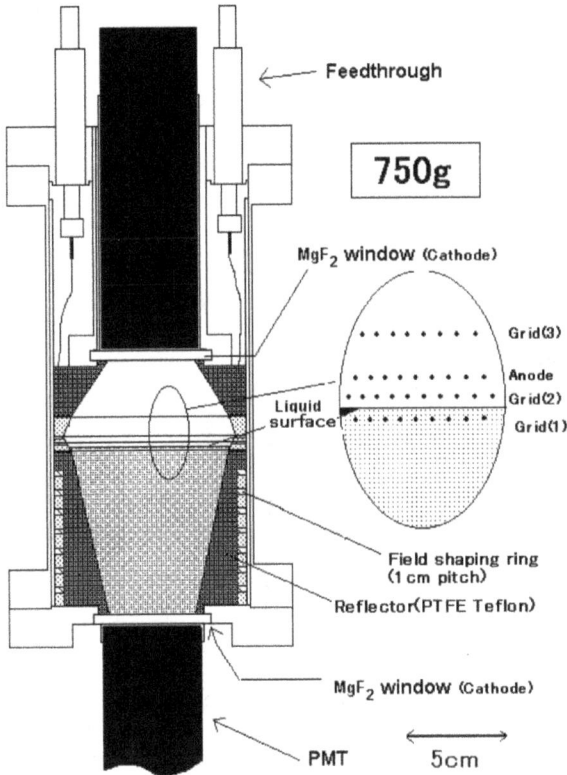

Figure 6.12 LXe time-projection emission chamber with two photo-multiplier readout (Suzuki *et al.*, 2000). Courtesy of M. Yamashita.

The two PMTs summarised typical γ-ray signal is shown in Fig. 6.13. The first small, narrow pulse corresponds to the scintillation flash generated at the moment of the γ-ray absorption in LXe. The second pulse is the electroluminescence flash generated by the cloud of ionization electrons drifting between the grid and the anode. The authors concluded that analyses of correlation between S1 direct (scintillation) and S2 proportional (electroluminescence) signals allows discrimination better than 99% between neutron recoils and gamma rays in the 10-100 keV range. A dependence of the measured energy resolution of the detector on γ-ray energy is fit by the function

$$\frac{\sigma}{E} = \frac{118}{\sqrt{E}} + 2.2\%$$ (6.1)

where E is measured in keV.

5 μs

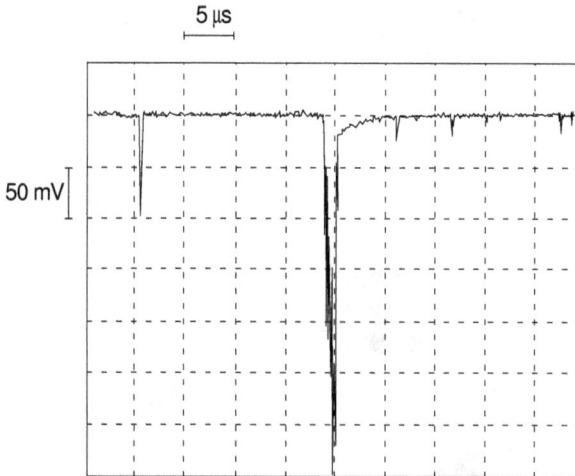

Figure 6.13 Waveform of a typical scintillation and electro-luminescence signals from LXe time-projection chamber shown in Fig. 6.10 at 250V/cm cathode-grid and 7 kV/cm anode-grid electric fields. Redrawn from Suzuki *et al.*, 2000.

Multichannel LXe TPC

The UK Dark Matter Collaboration in cooperation with US, EU and Russian collaborators is developing a series of detectors based on liquid xenon (Smith 2003). The first stage detector, ZEPLIN-I with fiducial mass of 3.1 kg, utilised only the scintillation light produced in particle interactions. The second generation of ZEPLIN detectors used emission detector technology detecting the both ionization and scintillation signals. These detectors are ZEPLIN-II (30 kg LXe) and ZEPLIN-III (6 kg LXe) and were used in the search for cold dark matter with expected sensitivity in WIMP cross-section to be $< 10^{-7}$ pb (see Chapter 8).

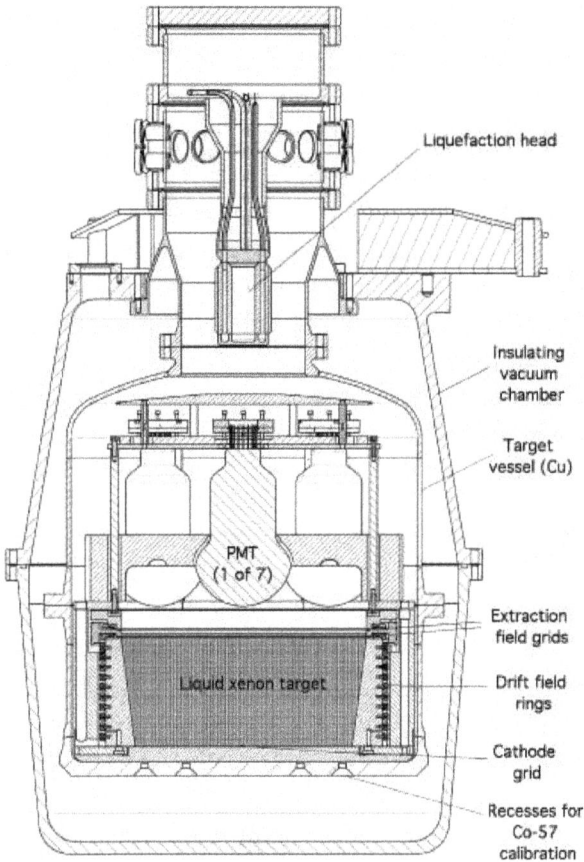

Liquefaction head

Insulating vacuum chamber

Target vessel (Cu)

PMT (1 of 7)

Extraction field grids

Liquid xenon target

Drift field rings

Cathode grid

Recesses for Co-57 calibration

Figure 6.14 Cut-away view of the ZEPLIN-II detector. Courtesy of T. Sumner.

Figure 6.14 shows a schematic drawing of the ZEPLIN-II detector (Alner *et al.*, 2007). A target mass of 30 kg is enclosed in a drift cage with PTFE conical reflecting walls viewed by an array of seven quartz-window 13 cm diameter ETL D742QKFLB photo-multipliers. The extraction region, where electrons are extracted from the liquid and the electroluminescence is generated, is defined by the two grids located on either side of the liquid surface. Xenon condensation occurs on the liquefaction head, with liquid dripping onto a copper shield which deflects it away from the photo-multiplier array and the active volume. A Compton veto system was provided for ZEPLIN-I LXe scintillation detector.

The next member of ZEPLIN family WIMP detectors, ZEPLIN-III (Fig. 8.3) has a shallow LXe drift volume without reflectors installed and because of the improved light-collection efficiency provided better discrimination between nuclear recoils and gamma rays. The ZEPLIN-III detector operates with an 8 kV/cm electric field in the liquid in order to record small ionization signals from the nuclear recoils as well as from the electron background. The active volume is comprised of a cylinder of 3.5 cm tall by 40 cm diameter and viewed by 31 PMTs. The PMTs record both scintillation and electroluminescence signals. ZEPLIN-III was used to get a physical result comparable to the best limit on WIMP cross-section measured with the XENON-10 emission detector as described in Chapter 7.

Multichannel LAr TPC

The European *WARP (WIMP Argon Program) collaboration* is developing liquid argon emission detectors for WIMP searches. This effort is based on the development of large LAr detectors for solar neutrino detection by the ICARUS collaboration (Amerio *et al.*, 2004). LAr may potentially provide better identification of nuclear recoils if time-analyses of decay times of the scintillation may be used as LAr contains two major components with very different decay times: $\tau_{singlet}$ = 7.0±0.1 ns and $\tau_{triplet}$ = 1.6±0.1 μs (Hitachi *et al.*, 1983).

In order to prove this approach, a prototype time-projection LAr emission detector was constructed as shown in Fig. 6.15. The sensitive volume of 1.87 L (2.6 kg mass) with a drift length of 7.5 cm is formed by focussing a PTFE reflector placed inside the field-shaping rings and reflecting cathode.

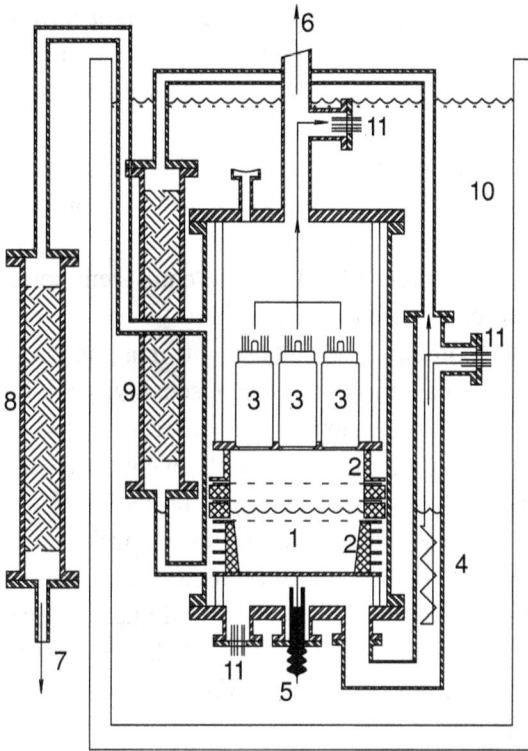

Figure 6.15 Layout of the WARP 2.6 kg LAr emission detector: liquid argon drift volume (1); reflector coated with wavelength shifter (2); 2" diameter photo-multiplier (3); resistor-heater (4); high-voltage feed-through (5); vacuum port (6); LAr inlet (7); filter-purifier for the detector filling (8); purifier for recirculation (9); LAr bath (10); multi-pin feedthroughs (11). Redrawn from Benetti *et al.*, 2008.

The detector was operated with a drift field of 1 kV/cm and electron-extraction and electroluminescence field of 4.4 kV/cm strength applied between grid electrodes. Seven 2"-diameter photo-multipliers were placed 4 cm above the last grid. Additional PTFE reflectors have been placed between the grids and the PMT array and the grids. The PMT windows and all reflecting surfaces were coated with wavelength shifter tetraphenyl butadiene (TPB) with quantum efficiency of conversion of argon VUV into blue visible range to be about 18%. The readout system was triggered by coincident 1.5 photo-electron signals from at least three PMTs that corresponded to 3.5 keV of energy deposited by nuclear recoil. The trigger efficiency is close to unite at generation more than 20 photo-electrons or 16 keV recoil energy.

The detector was immersed in a LAr bath supporting operation at 86.5 K. The purity of LAr was supported with continuous circulation of the gas through a getter. Residual concentration of electronegative impurities was of the order of < 1 ppb O_2 equivalent corresponding to a free electron drift path more than 0.5 m.

The detector combines advantages of imaging radiation field in electroluminescence cameras and position sensitivity in depth of working media in time-projection chambers. In order to exclude the effect of surface radioactivity of the reflectors, only events with drift times in the 10–35 μs range were considered useful. In respect to short-range nuclear particles, the detector was working as a wall-less detector (see Chapter 7). However, truly wall-less emission detectors can effectively reject γ-rays and β-particles from background nuclear decays that cannot be done effectively with a relatively small detector using liquid argon working medium with low stopping power for γ-rays.

This detector was used to measure scintillation conversion efficiency for argon nuclear recoils to be 1.26±0.15 phe/ keV in the 50–750 keV range and determined the γ-radiation rejection above the 35 photo-electron scintillation trigger to be better than $3 \cdot 10^{-7}$. The detector was tested in low-background experiment searching for cold dark matter conditions.

Table 6.1 presents the basic parameters of several electro-luminescence emission detectors with imaging capabilities for comparison. It is interesting to note that the energy resolution of the first two-dimensional imaging detector constructed by Egorov *et al.* (1983) still represents the best record on energy resolution among many emission detectors constructed for the past 25 years, probably because of better light-collection efficiency provided by large photo-multiplier array imaging a relatively thin working medium without use of any reflector.

Table 6.1 Properties of electroluminescent emission detectors.

Media	Size, cm	$\Delta E/E$, % FWHM (E keV)	Δx, Δy, mm	Δz, mm	Read-out	Ref.
S/G Xe	∅22·0.5	15 (122)	3.5		19 PMTs	a
L/G Xe	∅22·0.5	16 (122)			19 PMTs	a
S/G Kr	∅22·0.5		2.5		19 PMTs	b
L/G Xe	∅2·2			0.5	1 PMT	c
L/G Xe	∅5·1.5	24 (122)			1 PMT	d
L/G Xe	∅3.7·0.96	28			2 PMTs	f
L/G Xe	∅5·1.5	18 (50ee)		0.4	1 PMT	g
L/G Xe	∅6·1				1 PMT	h
L/G Ar	∅15·7.5				7 PMTs	l
L/G Xe	∅8.4·∅4.6·8.5	30 (122), calc.			2 PMTs	k
L/G Xe	∅40·3.5				7 PMTs	m

Notes: (PMT) photo-multiplier; (E) deposited energy; (L) liquid; (G) gas; (S) solid; (ee) electron-equivalent; (calc.) calculated value; Ref.: (a) Bolozdynya, Egorov *et al.*, 1982; Egorov *et al.*, 1987; (b) Bolozdynya, Egorov *et al.*, 1981; (c) Guschin *et al.* 1982; (d) T. Shutt, private communication; (f) Shutt *et al.*, 2006; (g) Afanasiev *et al.*, 2003; (h) Aprile, Giboni *et al.*, 2004; (l) Amerio *et al.*, 2004; (k) Suzuki *et al.*, 2005; (m) Smith, 2003.

Chapter 7

Emission Detectors for Low-background Experiments

Emission detectors are electron drift devices that operate by collecting ionization electrons from condensed phase target media. Due to long charge collection times, they are relatively slow and inconvenient for use in modern high-count rate accelerator experiments. However, they are a unique tool for low-background experiments searching for rare events. Examples include interactions of cold dark matter with normal matter, neutrino scattering, and double-beta decay.

7.1 Wall-less emission detectors

A primary challenge for experiments searching for neutrinoless double-beta decay, for solar neutrino flux from the pp-cycle, and for cold dark matter is the need to reduce the environmental beta and gamma backgrounds. In some cases, where detectors require little material for containment, the use of metal shielding with a low concentration of radioactive impurities reduces the background to a level that does not limit currently achievable sensitivities, for example, in the neutrinoless double-beta decay experiments, down to a level of $T_{1/2}^{0\nu}(^{76}Ge) \sim 10^{26}$ yr. In the case of cryogenic or high-pressure gas detectors, radioactive impurities within materials used to construct massive vessels often limit sensitivities of low background experiments (see, for example, Ovchinnikov and Parusov, 2002).

Due to the high sensitivity of photo-multipliers, emission detectors based on pure noble gases allow for the detection of both excitations and ionizations and the achievement of three-dimensional position sensitivity that can be used to operate as *'wall-less'* detectors for rare and low-ionizing events (Bolozdynya *et al.*, 1995).

The wall-less detector works as follows (Bolozdynya, 1999):

1) Radiation interacts with the condensed target medium, exciting and ionizing atoms; this process generates a prompt signal that manifests itself in the form of scintillation in noble liquids and solids, phonons in crystals, and rotons in superfluid helium. This signal serves as a trigger.

2) In response to the applied external electric field, ionization electrons drift to the interface, escape into the rarefied gas or vacuum region (or superconductive collector, for crystal targets) and generate a second, amplified signal. Different processes can be used for signal amplification: electroluminescence of the gas phase, electron avalanche

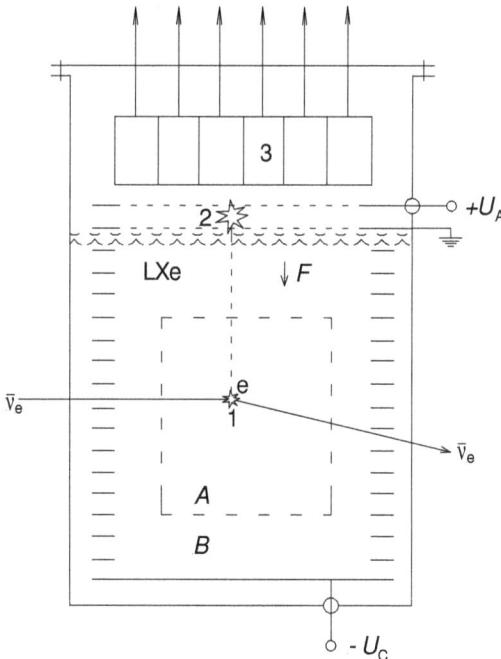

Figure 7.1 Principle of operation of a wall-less emission detector: scintillation flash at the point of interaction (1); electroluminescence flash generated by electrons extracted from the liquid and drifting through the gas (2) array of photo-electron multiplier tubes (3); (A) fiducial volume; (B) shielding layer. Redrawn from Bolozdynya *et al.*, 1995.

multiplication in low density gas, acceleration of electrons in vacuum, the Trofimov–Neganov–Luke effect in crystals, breaking *Cooper pairs* and generation of a pulse of quasi-particles in superconductors, etc. An array of sensors is used to measure the two-dimensional distribution of the secondary particles and to determine the coordinates of the original event on the plane of the sensor array. Since the second signal is delayed from the first one, the third coordinate of original interaction is also uniquely determined.

3) From the three-dimensional position reconstruction, a *fiducial volume* can be defined (A, Fig. 7.1). Events originating in the vicinity to the detector walls can be eliminated as being potentially associated with radioactive background radiated from the surrounding materials. By making the detector sufficiently large and choosing a target medium with a high stopping power for nuclear radiation, the fiducial volume is effectively shielded by the outer detector medium layer (B, Fig. 7.1).

It is important to point out that there are other detector technologies that can be used to construct 'wall-less' detectors. For example, bulk scintillators viewed by a photo-detector array totally surrounding the 'crystal-ball' have been considered as 'wall-less' detectors for such experiments as XMASS, CLEAN. However, emission detectors based on pure noble gases require only a few readout channels: the first signal is proportional to the excitation of the condensed medium; the second is proportional to the ionization. Since the efficiency of different modes of dissipation of the deposited energy depends on the nature of interactions, multi-mode readout helps distinguish events of different origin and effectively suppresses the background (Cline, 2000; Aprile, 2002; Smith, 2003). In superfluid helium, in addition to photons and electrons such quasi-particles as rotons can be generated. In solid noble gases, phonons may be observed at low temperatures. These features, along with the availability of super-pure noble gases in large amounts, make condensed noble gases the most attractive media for emission detectors of rare events.

Xenon is a noble gas from which electronegative impurities can be readily removed. It has no long-living radioactive isotopes and, thus, the intrinsic radioactivity of a xenon target can be reduced to a very low level. Due to its high stopping power for γ-radiation, liquid xenon emission detectors are the most attractive media for construction of massive 'wall-less' self-shielded detectors that have become a powerful new technology for the detection of dark matter (Chapter 8). Figure 7.2 shows the shielding effect of pure liquid xenon for a few hypothetical liquid xenon (LXe) emission detectors with a total mass of xenon ranging from 10 to 1000 kg and placed in realistic background conditions of XENON-10 experiment. Depending on the selected fiducial volume, different levels of suppression of the natural background are achievable. For example, selecting a 20 kg fiducial 'core' in a 100 kg LXe detector allows a background suppression factor of 30; selecting a 100 kg 'core' as the fiducial volume inside a 1 tonne LXe detector allows a background suppression factor of 300.

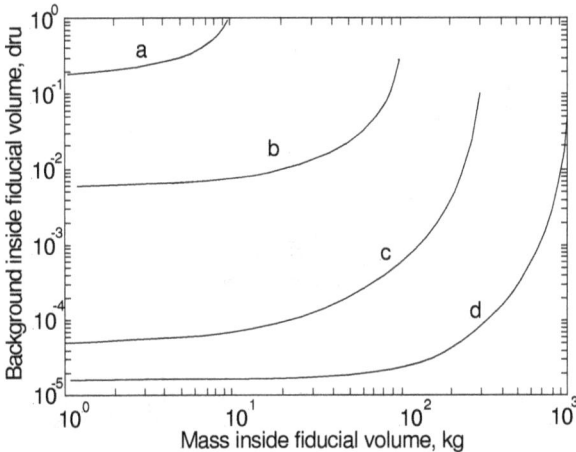

Figure 7.2 Self-shielding effect in LXe emission detectors. The calculated background inside detectors of total active mass 10 (a), 100 (b), 300 (c), and 1000 (d) kg LXe as a function of the mass within a given fiducial volume. The background unit 'dru' is 1 event/kg/keV/day. Courtesy of R. Gaitskell and L DeViveiros of LUX collaboration.

7.1.1 XENON-10 and XENON-100 detectors

The first truly 'wall-less' emission detector was constructed by the XENON collaboration for a cold dark matter search. The detector, XENON-10, contained the active xenon target of 13.5 kg mass enclosed inside a cylindrical Teflon UV diffusive reflector and copper drift rings cage for electric field shaping (Fig. 7.3). The structure of the drift cage was designed for a maximum field of 5 kV/cm and a 15 cm LXe drift gap. Two arrays of 1″ square metal envelope photo-multipliers (Hamamatsu R8520-AL), 48 in the gas (Fig. 3.13) and 32 in the liquid, were used to detect both primary scintillation of the liquid (prompt signal, referred to as S1) and electroluminescence of the gas (delayed signal, referred to as S2). The detector was enclosed in a cylindrical cryostat ($\varnothing 70 \cdot 100 \text{ cm}^2$ dimensions) and surrounded with passive shielding.

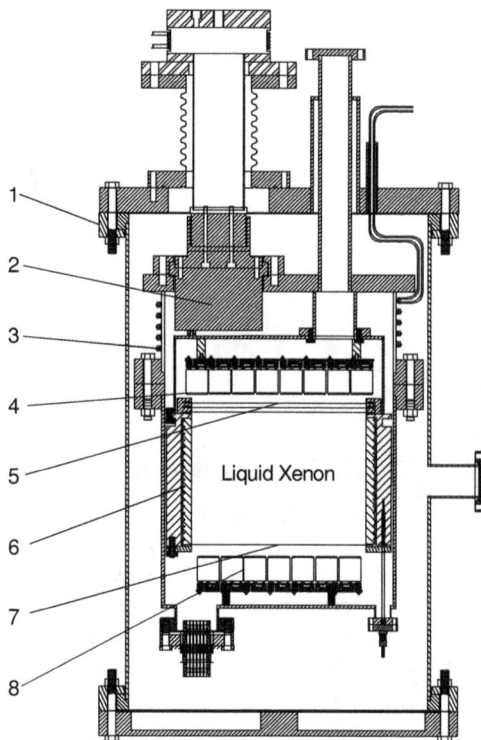

Figure 7.3 Schematic view of XENON-10 liquid xenon emission detector: vacuum vessel (1); refrigerator (2); cooling coil (3); top array of 48 photo-multipliers (4); grids (5); drift electrodes and Teflon reflector (6); cathode mesh (7); bottom array of 32 photo-multipliers (8) (Aprile, 2005).

The elastic scattering of a WIMP with a liquid xenon nucleus is expected to result in low-energy Xe recoil. The recoiling Xe atom produces both ionization electrons and fast UV scintillation photons at 175 nm, from the de-excitation to the ground state of excited diatomic xenon molecules (Xe_2). Under a high electric field, the nuclear recoil will yield a very small charge signal and a much larger light signal, compared to electron recoil of the same energy. The distinct charge/light ratio is the basis for nuclear recoil vs. electron recoil (background) discrimination in a LXe detector. To detect the small (a few photo-electrons) signals involved, the process of electroluminescence is used. Ionization electrons that escape recombination are extracted from the liquid to the gas phase, where in a strong electric field they induce proportional scintillation light. The number of photons generated by one drifting electron is sufficiently large to be detected by PMTs.

The XENON-10 detector was commissioned at Laboratori Nazionali del Gran Sasso (LNGS) in 2006. The PMT signals (48 from the top array, 41 from the bottom array) showing S1 and S2 for a typical 5 keVee electron recoil (ER) event are shown in Fig. 7.4. The hit pattern on the top PMTs (S2 signal) allows the reconstruction of the *xy* position, while the drift time (delay between S2 and S1 signals) gives the *z* (depth) coordinate of the point-like event. With an applied drift field of 0.72 kV/cm over the drift length of 15 cm, S1 signals have an average yield of 2.2 (3.0) phe/keVee based on the 662 (122) keV gamma peak, from ^{137}Cs (^{57}Co) calibrations. For an event in the middle of the detector 80% of the S1 signal appears in the bottom PMTs, and 20% in the top – primarily due to total internal reflection at the liquid–gas interface. The lifetime of the drift electrons (S2 signal) steadily improved during the first few weeks of recirculation and was confirmed to be in excess of 2 ms (~ 4 m electron drift length) for the duration of the dark matter search. The S2 yields ~ 450 phe/keVee, and it was also established that one electron, when extracted into the gas, generates ~ 24 phe in the PMT array (specific to the operating temperature, pressure and gas drift gap).

A similar performance but with factor 10 better sensitivity to WIMPs is expected for a new XENON-100 detector (Fig. 7.5) installed at the Gran Sasso Underground Laboratory in 2008. The detector contains of 170 kg LXe with a likely 65 kg fiducial mass, and is viewed by 242

R8520-06-AL PMTs. The active target is enclosed in a light-collecting PTFE reflector of 30 cm diameter and 30 cm height. With a 6000 kg·day exposure, XENON-100 will expects to reach a sensitivity to spin-independent WIMP-nucleon cross-section of $2 \cdot 10^{-45}$ cm^2 (Aprile and Baudis, 2008).

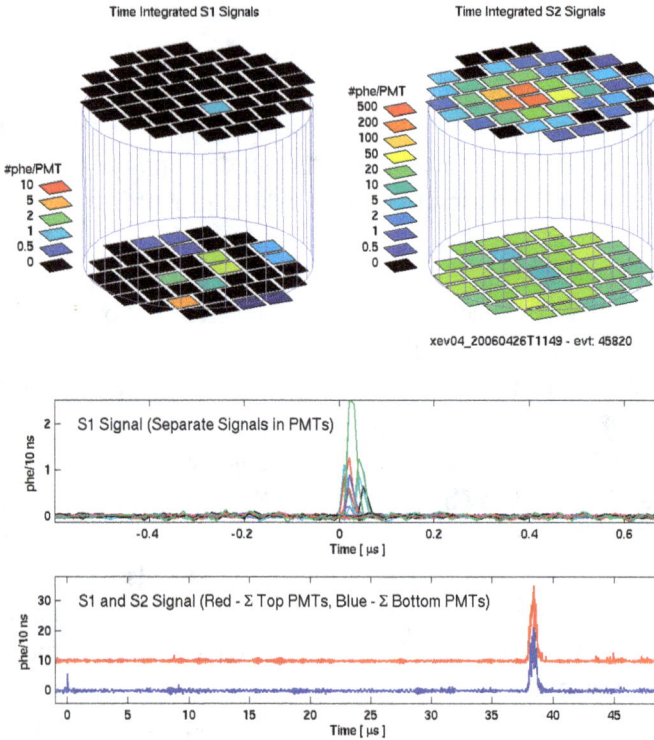

Figure 7.4 Typical distribution of individual PMT signals over the top and the bottom PMT arrays of XENON-10 detector for 5 keVee event used to form S1 and S2 signals. Courtesy of E. Aprile and L. DeViveiros of XENON-10 collaboration.

In the upgraded XENON-100 detector, the bottom array of PMTs may be replaced with 19 QUPID (quartz photon intensifying detectors) low-radioactive photo-sensors (Arisaka *et al.*, 2008). The stainless steel vessel will be replaced with a low radioactive one made of oxygen-free copper (OFHC). To reduce radioactive background from the top array of

98 PMTs that would then dominate the drift depth of the detector will be increased from 30 cm to 60 cm. An upgraded XENON-100 will improve its sensitivity by another order of magnitude by 2012.

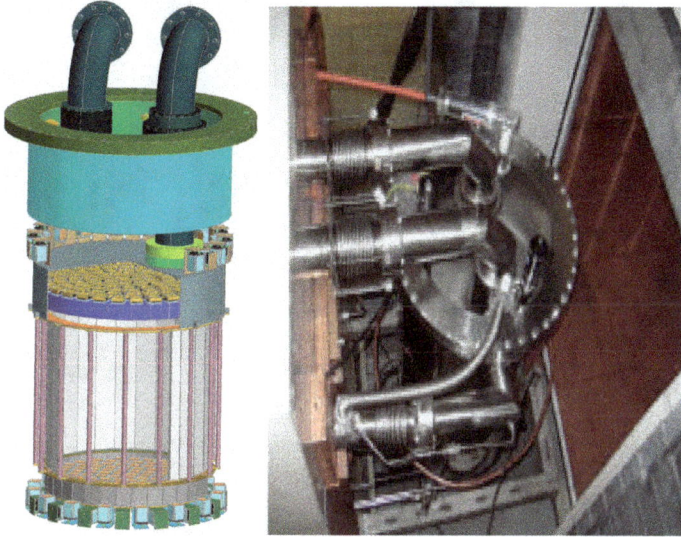

Figure 7.5 Schematic view (left) and a picture (right) of XENON-100 emission detector installed inside passive shield. Courtesy of E. Aprile.

7.1.2 LUX detector

The Large Underground Xenon (LUX) collaboration is developing a 300 kg active mass, two-phase LXe emission detector for a deep underground experiment that will extend current dark matter sensitivity two orders of magnitude beyond current best limits, to an event rate better than ~ 1 event/100 kg/month. This corresponds to a spin-independent WIMP-nucleon cross-section of $7 \cdot 10^{-46}$ cm^2. A schematic drawing of the detector is shown in Fig. 7.6. The detector consists of two enclosed vessels made of low-background titanium. The internal vessel is surrounded with a copper thermal screen and supported at 165–190 K temperature with thermosyphons (Bolozdynya *et al.*, 2008) installed onto the thermal screen.

Figure 7.6 A prototype of the LUX detector tested at CWRU in 2008–2009 (left) and a schematic drawing of the LUX detector in titanium cryostat (right). Courtesy of the LUX collaboration.

The LUX detector will use two arrays of sixty Ø2.2" R8778 (Hamamatsu) PMTs to measure the active Xe region. One, above the liquid surface is primarily used to image the xy-position of the S2 light pulse with about 1 cm position accuracy at low energy. The second array is in the liquid, below the cathode grid. The Hamamatsu R8778 photo-multiplier, specifically developed to operate in LXe at ~ 170 K, is able to tolerate up to 4 bar pressure without risk of implosion and has been extensively tested by the XMASS collaboration. The quartz window and photo-cathode (PC) material are optimised to achieve quantum efficiency (QE) of > 30% at 175 nm of Xe, and the collection efficiency for electrons from the PC onto the first dynode is ~ 90%. The combination of these two factors gives an effective QE of ~ 27%.

The total light collection depends on the effective QE of the PMTs, the reflectance of the PTFE and grids, and on any absorption in the liquid. Comparison of simulations and calibration data of the (x, y, z) variation in S1 light collection in XENON-10 obtain best fits with PTFE reflectance at 98%. Because the PTFE is so highly reflective, Rayleigh scattering is not important: scattering length $\lambda_R = 40\text{--}50$ cm obtained experimentally (Baldini *et al.*, 2005), 30 cm from theory (Seidel *et al.*, 2002). Self-absorption of Xe scintillation light should be absent except on impurities because of the mechanism of its production from decays of self-trapped exciton states (Doke, 1981). Recent MEG results (Baldini *et al.*, 2005) find $\lambda_{abs} > 1$ m at 90% CL, for impurities (primarily O_2 and H_2O) < 100 ppb.

7.1.3 Next generation of wall-less emission detectors

Using a xenon mass of 1 tonne (about 70 cm linear size of the sensitive volume), a detector can reach a sensitivity below 10^{-46} cm^2 for spin-independent cross-sections, which is several orders of magnitude below the best current limit (Aprile *et al.*, 2005).

In LXe emission detector with total mass > 10 tonne, the suppression of the radioactive background due to self-shielding effect and multiple vertex events suppression can be so high that neutrino interactions from the sun become the major background factor limiting the sensitivity of the detector to WIMPs (Monroe and Fisher, 2007). Scattering of pp-solar neutrinos would be expected as a flat component in the energy range 0–50 keVee, with an event rate of $1.2 \cdot 10^{-5}$ events/keVee/kg/day (Fig. 8.4).

The next generation of wall-less emission detectors are expected to contain 20–30 tonnes of LXe totally surrounded by photo-detectors, as shown in Fig. 7.7. Such detectors are foreseen to be used for multiple tasks, including WIMP searches, detection of neutrinos from the sun, and the search for neutrinoless double-beta decay (Arisaka *et al.*, 2009).

7.2 Identification of radiation in emission detectors

There are a few methods for the identification of particles detected in emission detectors. Those methods are based on potential three-dimensional position sensitivity, multi-signal response and the large mass and size of detectors. Identification capability along with the selection of events in an inner fiducial volume is a powerful tool for background rejection, in the context of searching for extremely rare events.

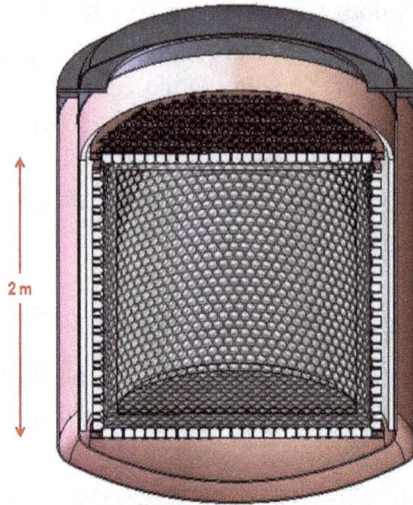

Figure 7.7 Project of the XAX detector as described in Arisaka *et al.*, 2009. Courtesy of K. Arisaka.

7.2.1 *Scintillation signal wave-form analyses*

The scintillation of noble gases is a complicated process involving the de-excitation of different excitations of the target atoms. The decay times and intensities depend on excitation and recombination processes, and on the nature of the detected particle. Analyses of the *kinetics of scintillation* in certain cases can be used for identification of the detected radiation. In noble liquids and solids, to first approximation, the intensity of scintillation may be expressed as a superposition of a few independent exponential terms:

$$i(t) = \sum_{i=1}^{n} a_i \cdot \exp(-t / \tau_i) \qquad (7.1)$$

If the relative intensity of each component, a_i, does not depend on time (that is not true in some special cases such as for liquid neon and for combination of noble scintillators with solid wavelength shifters — see monograph, Aprile *et al.*, 2006 for details), the integral intensity of each component is simply the value of the coefficient, a_i.

In the most popular media of large emission detectors — liquid argon and liquid xenon — the kinetics of scintillation may be described with formula (Eq. 7.1), with the number of components set as low as $n = 2$. Two major components of scintillation in these media are associated with singlet and triplet exciton states. In liquid argon, scintillation can be described with $\tau_s = 7.0\pm1.0$ ns and $\tau_t = 1600\pm$ 100 ns — and their relative intensities a_s / a_t — are found to be 0.3, 1.3 and 3 for electron, α-particle and ion recoil excitations, respectively (Hitachi *et al.*, 1983). In liquid xenon, $\tau_s \approx 3-4$ ns and $\tau_t \approx 20-30$ ns — and their relative intensities a_s / a_t — are found to be ~ 0.5 and ~ 2 for α-particle and ion recoil excitations; for electrons, only the triplet exciton decay has been observed. Thus, in the both LAr and LXe emission detectors, the kinetics of scintillation may be used to distinguish between electron and recoil excitations. However, the efficiency of this method strictly depends on the efficiency of light collection. For low-energy recoils, a very weak scintillation signal is expected, and this limits the utility of this method.

7.2.2 Multi-response analyses

Emission detectors using noble working media can record two signals originating from the excitation and ionization of target atoms, on an event-by-event basis. Different channels of energy dissipation are distinctly sensitive to the ionization density and nature of the particle interaction (i.e. electrons interact via the electro-weak force; nuclei interact mostly via the strong force). Hitachi, Yunoki and Doke (1987) have shown that, under the application of an electric field, the ionization yields increase while and excitation yields decrease as the applied field strength is increased. Moreover, they have shown that for different particles (alpha or fission fragments) the dependences of the yields on the electric field behave differently. This observation leads to the idea of using the ratio of S1-to-S2 signals in emission detectors, for distinguishing between electrons (the major background in small and medium-size detectors) and the nuclear recoils that could be an indication of a WIMP scatter.

For liquid xenon, electron recoil *background discrimination* based on light and charge has been characterised in prototype detectors by several groups (Yamashita, 2004; Aprile *et al.*, 2006; Shutt *et al.*, 2006), and used for analyses of data acquired by the ZEPLIN-II, ZEPLIN-III and XENON-10 dark matter detectors. Dense nuclear recoil tracks have lower charge yield than less-dense electron recoils, so that the S2/S1 ratio bands of each are distinct. The charge yield for nuclear recoils increases as the energy decreases, consistent with the fact that the electronic stopping power of nuclear recoils (and hence the charge density) drops rapidly with decreasing energy in this regime. Below 20 keVr the charge yield for electron recoils increases precipitously.

The XENON-10 analysis has been based on 4.5 keVr detection threshold for nuclear recoils; at this energy, the measured discrimination efficiency was 99.9%, decreasing to 99.3% at 27 keVr. Further, the electric field dependence of the charge yields is nearly absent for nuclear recoils, and is weak for low energy electron recoils, so that discrimination is essentially independent of electric field (Fig. 7.9). The discrimination is limited by the width of the electron recoil distribution, which has been shown (Shutt *et al.*, 2006) to be dominated primarily by fluctuations in the recombination. Statistical fluctuations in the scintillation (S1) light statistics are present, but subdominant above ~ 5 keVr.

7.2.3 Analysing topology of events

Already the first true imaging emission detector — emission streamer chamber — has been considered as an instrument for identification of extremely low-ionizing particles, tracks of which are consisted from single electrons generated in condensed medium with density < 1 cm^{-1} (Bolozdynya *et al.*, 1980).

Emission detectors with digital readout can easily identify *multi-vertex events*: point-like events as well as long tracks of relativistic particles in condensed matter. According to simulations provided by the LUX collaboration, the neutron background for this experiment could be suppressed by a factor of 20 by rejecting multiple vertex events. These would occur due to multiple neutron elastic scattering, before the neutron either left the detector target or capture in detector material.

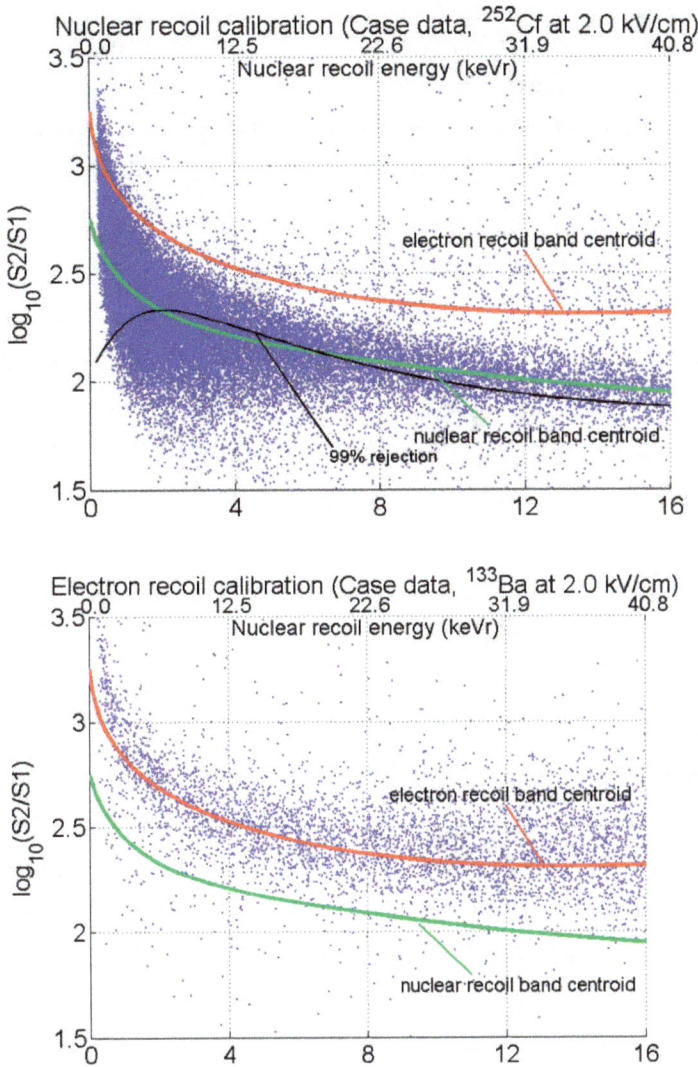

Figure 7.9 $\log_{10}(S_2/S_1)$ factor as a function of the energy (keV) of electron recoils (top) and nuclear recoils (bottom) for low energy range of deposited energies due to 662 keV (^{137}Cs) γ-rays interactions and neutron scattering from AmBe neutron source in liquid xenon: red line is mean of distribution and nuclear recoils (top), green line is mean, along with 99% and 99.9% ER discrimination levels, black points, deduced from the data (Shutt *et al.*, 2006).

Unusual multi-vertex events — five, coplanar or point-like ionization depositions — can be a unique signature of positron double-beta decay as described in Section 8.3.1.

Thus, the imaging capability of emission detectors with bulk working medium suitable for 3D imaging of interactions of highly penetrating radiation makes them an important tool for low-background experiments of fundamental importance.

Chapter 8

Applications of Emission Detectors

Near the end of the 19th century, it was discovered that atoms were not the fundamental particles of nature, but conglomerates of smaller particles. The early 20th century saw explorations of nuclear decay and nuclear reactions that, by 1939, culminated in the experimental discovery of nuclear fission and the theoretical explanation of how nuclear fusion powers the sun. These developments demonstrated that atomic nuclei are also composed from smaller particles. Soon, new particles were found in cosmic ray experiments. The many scattering experiments that were carried out after the development of accelerators of electrons, protons and later anti-particles led to the discovery of a 'particle zoo' that compelled physicists to formulate the Standard Model during the 1970s, in which the variety of discovered particles was explained as combinations of a relatively small number of fundamental particles.

By the end 20th century, particle physics focussed on discovering what may lie beyond the Standard Model. The development of the most powerful accelerators, colliders of massive ions, such as the Large Hadron Collider, is one of the two major ways to explore a new hypothetic world of supersymmeric particles.

Another powerful method to explore this very high energy region is by non-collider and low-background experiments searching for dark matter, neutrino mass and proton decay. Such experiments utilise massive, highly sensitive detectors that are well-shielded from natural radioactive sources and very often work deep underground.

At the beginning of the 21st century, underground high-energy neutrino detectors, such as Super-Kamiokande, SNO, and KamLAND, proved that neutrinos have mass. The discovery of neutrino mass became the first substantial change in the Standard Model of particle physics in the last 20 years. In addition, new, highly sensitive, dark matter detectors constructed by the CDMS, EDELWEISS, ZEPLIN, and XENON collaborations have driven down the limits for cold dark matter scattering cross-sections. In this race to discover new physics beyond the Standard Model, emission detectors are taking a leading role today. Emission detector technology has become the basis for the next generation of the cold dark matter experiments searching for WIMPs. Emission detectors are also being considered for the detection of rare events such as low-energy neutrino interactions and double-beta decay. All these exciting opportunities look achievable, because emission detectors have an impressive combination of properties:

- extremely effective suppression of the natural radioactive background due to three-dimensional imaging capability with electronic readout and effective self-shielding;
- availability in huge masses (hundred tonnes) in order to provide a reasonable counting rate for events with extremely low cross-sections;
- capability for additional rejection of background and identification of particles due to multi-mode readout and waveform analyses of acquired excitation and ionization.

In this Chapter, we consider a few fundamental experimental tasks that can be solved using emission detectors.

8.1 Detection of weakly ionizing particles

Tracks of cosmic rays discovered by C.T.R. Wilson with his *cloud chamber* inspired the development of 'the most original and wonderful instrument in scientific history', according to Lord Rutherford, and led to the Noble Prize for the inventor in 1927 (for references see, for example, Boors and Motz, 1966). The later developments of diffusion, bubble and spark chambers provided further opportunities for visualising particle interactions. These approaches have been used since that time for the exploration of the most fundamental principles of particle physics. The very first experiments with tracking detectors demonstrated that some particles have reduced ionization power, and the detection of beta-particles, for example, was considered as a special technical task for a while (Spence *et al.*, 1948). At about that time, the investigation into the penetration of high energy cosmic rays deep underground began, and that fuelled the interest in the weakly ionizing components of cosmic radiation (Miezowicz *et al.*, 1950). Attempts to detect relativistic particles with fractional charge have been undertaken since the discovery of quarks in e^+e^- annihilations (Aihara *et al.*, 1984). By the end of 20th century, interest in the detection of weakly ionizing particles grew when it was recognised that such particles might compose the majority of matter in the universe.

8.1.1 Tracking particles with minimal ionization power

After the ability of emission chambers to detect single electrons was demonstrated, the idea of using an emission streamer chamber to search for particles with extremely low ionizing power was formulated by B.U. Rodionov. The idea was based on the fact that an easily detectable spark discharge (streamer), visible even by the naked eye, can be initiated by a single electron. A few streamers detected simultaneously and located along a straight line may serve as evidence that they are generated by the same weakly ionizing particle. The first experiments with an emission streamer chamber in 1977 (see Section 6.1.2) showed that, indeed, detected tracks sometimes look like a chain of streamers distributed along a line with a density of ~ 1 cm^{-1} that is, at least, an order of value smaller than the density of streamers distributed along tracks of relativistic pions (Bolozdynya, Egorov *et al.*, 1980). Such 'abnormal' tracks comprised about 3% of the population of tracks detected with the emission streamer chamber, triggered with a telescope of scintillation detectors positioned at the secondary beam of the ITEP proton synchrotron. Analysing possible mechanisms imitating the 'abnormal' tracks, the authors concluded that this is a 'memory' effect associated with ionization electrons captured on the solid krypton surface. They found that 'abnormal tracks' are, unusually, often generated exactly at the same place where the previous track was detected and often accompanied by a 'normal' track. It is now understood that some fraction of ionization electrons trapped on the interface during the first detection act are liberated by high voltage pulse and detected together with the next, closely following, triggered event.

One of the possible mechanisms considered to explain this phenomenon may involve photo-electron emission from negative ions located under the surface as a potential image of an already detected track. The ions can be produced by the capture of low-energy, non-emitted electrons by electronegative impurities such as oxygen. The intensive flash of UV light ($\sim$ 7 eV energy of photons) accompanying the streamer discharge may liberate electrons and allow some of them to escape into the gas phase where they can be visualised as a rare chain of streamers.

Special experiments with an X-ray generator demonstrated that, indeed, the emission streamer chamber had a memory of about 1–2 µs, corresponding to the free-electron life time in the solid krypton used in this experiment. Later experiments with the XENON-10 emission detector indicated that there is a delayed single-electron emission associated with the random escape of electrons trapped under the interface of the condensed noble phase. The random emission such trapped electrons may explain single-electron afterpulses observed in other experiments with emission chambers (see, for example, afterpulses observed in the XMASSII detector as shown in Fig. 6.13).

8.1.2 Detecting cold dark matter

The most fundamental problem of modern astrophysics is the missing mass (*dark matter*) of the universe, which indicates its presence only via gravitational forces. Dark matter comprises a majority of the gravitating mass ($\sim$ 25%) with a local density between 0.4 and 0.7 $(GeV/c^2)/cm^3$ and an average velocity around 230 km/s (Lewin and Smith, 1996), yet, the nature of dark matter remains unknown, providing a central problem for cosmology for more than two decades. One popular explanation of the invisible mass is that it consists of *weakly interacting massive particles* (WIMPs) expected in the Supersymmetry model. The model predicts that the mass of such particles is in the range of 10 to 10^4 GeV/c^2 and that the cross-section lies in the range of 10^{-42} to 10^{-48} cm^2 (Ellis *et al.*, 2005).

In 1989, Barabash and Bolozdynya suggested using an emission detector filled with ~ 100 kg of saturated hydrocarbon liquid (iso-octane) at room temperature to search for WIMPs of 1–10 GeV/c^2 mass (Barabash and Bolozdynya, 1989). Figure 8.1 shows a schematic drawing from that proposal. Such a detector should have little sensitivity to the natural γ-ray background, and it would operate at room temperature without the need for a complicated and massive cryostat. Since that time, most CERN collider experiments have excluded masses below 80 GeV/c^2 and cross-sections above ~ 10^{-42} cm^2 that were predicted by minimal supersymmetric models (Ellis *et al.*, 2005). There are a number of recent experimental and theoretical results, which limit the parametric space of cross-section-vs.-mass that should be checked by upcoming experiments discussed in this section (Fig. 8.2).

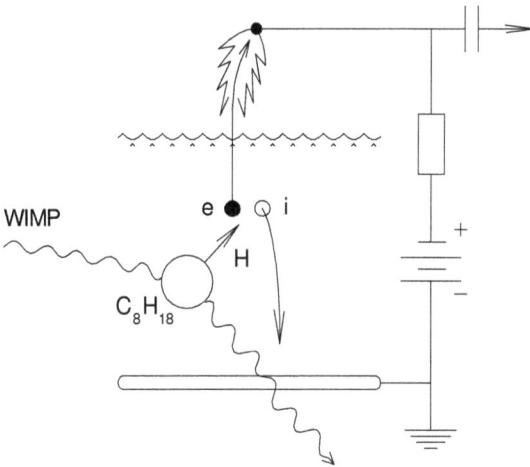

Figure 8.1 Principle of operation of 100 kg liquid iso-octane emission detector proposed for detection of WIMPs with 1 GeV/c^2 mass. Redrawn from Barabash and Bolozdynya, 1989.

In the course of the development of large noble liquid detectors, it was realised that large emission detectors, triggered by scintillation signals and detecting both scintillation and electroluminescence signals with arrays of photo-detectors, can operate as wall-less detectors (Bolozdynya *et al.*, 1995). Through the selection of events inside the

fiducial volume of the large emission detector and by using the outer
layer of the pure noble liquid as a radiation shield, emission detectors can
provide extremely high sensitivity for low-background experiments
(Fig. 7.1). Modern purification technology of noble gases provides more
than 1 m drift length of quasi-free electrons and ~ 1 m attenuation length
for scintillation photons in condensed xenon. This means that detectors
with a multi-tonne working mass could be constructed (Bolozdynya,
1999), therefore making liquid–gas xenon emission detectors the most
promising technology for search of WIMPs with mass around
100 GeV/c^2 (Fig. 8.2).

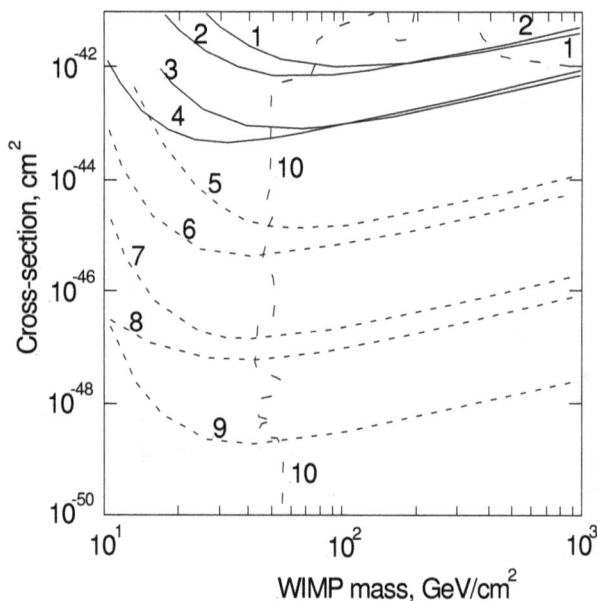

Figure 8.2 The elastic scattering cross-section for spin-independent couplings
versus WIMP mass, including experimental data from WARP 2.3L (1);
ZEPLIN-II (2); ZEPLIN-III (3); XENON-10 (4) experiments and projections
for next generation of emission detectors of LUX 300 kg (5); 3 t (7), 20 t (9)
and XENON-100 (6) and 1 t (8) in comparison with theoretical predictions
inside the contour (10). Redrawn from the graph automatically generated at
the Web-plotter http://dendera.berkeley.edu/plotter/entryform.html maintained
by R. Gaitskell, V. Mandic and J. Filippini.

In 2007, the best limits for WIMP–nucleon cross-sections were reported from a 136 kg·d exposure of the XENON-10 LXe emission detector at the Gran Sasso Underground Lab (Angle *et al.*, 2008a, b). At the beginning of 2009, the best limit for the spin-dependent WIMP–nucleon cross-sections was reported from the first run of the ZEPLIN-III LXe emission detector (Lebedenko *et al.*, 2009).

To improve sensitivity to different rare processes, the next generation experiments being considered will use multi-tonne emission detectors utilising different target nuclei and background discrimination methods (Arisaka *et al.*, 2009).

XENON experiment

XENON-10 is the first stage of the XENON experiment intended to directly detect weakly interacting massive particles (WIMPs). With the XENON-10 detector, described in Section 7.1.1, installed at the Laboratori Nazionali del Gran Sasso (LNGS), low-background data were taken from December 2006 through February 2007 at a WIMP mass of 30 GeV.

Due to the relatively small size of XENON-10 the background reduction due to active self-shielding was relatively modest. The background was further suppressed by passive gamma and neutron shielding and the detector's three-dimensional position sensitivity. The passive shield has 30 cm of polyethylene (2.2 tonnes), and 23 cm of Pb (27 tonnes). The electron-recoil (ER) background in an inner fiducial volume (FV) of 5.4 kg LXe was ~ 0.6 evt/keVee/kg/day[*] for single scatter events with valid S1 and S2 signals. XENON-10 clearly demonstrated that, by selecting an inner FV, one can effectively eliminate event with partial charge collection when it occurs near the edge of the Xe active volume. After cuts were applied to remove anomalous events, the energy window of interest has been analysed for the 58.6 live-days of WIMP-search data. At the 90% CL, the upper limit

---

[*] The unit keVee is used for electron recoil equivalent energy; keVr is used for nuclear recoil equivalent energy.

for the WIMP-nucleon cross-section was determined to be $8.8 \cdot 10^{-44}$ cm^2 at a WIMP mass of 100 GeV and $4.5 \cdot 10^{-44}$ cm^2.

New results for spin-dependent WIMP-nucleon interactions with ^{129}Xe and ^{131}Xe were reported in 2008 (Angle *et al.*, 2008) after 58.6 live days of operation. Based on the non-observation of a WIMP signal in 5.4 kg of fiducial liquid xenon mass, previously unexplored regions in the theoretically allowed parameter space for neutralinos were excluded. Also excluded was a heavy Majorana neutrino with a mass in the range of ~ 10 GeV/c^2–2 TeV/c^2 as a dark matter candidate under standard assumptions for its density and distribution in the galactic halo.

Much greater sensitivity will be achieved in the next 100 kg active target experiment, named XENON-100, currently running at LNGS. This member of the XENON detector family (Fig. 7.5) contains 170 kg of xenon with 65 kg in the FV and 105 kg in the active shield. With a 6000 kg·d exposure, the experiment is expected to reach a sensitivity for spin-independent WIMP-nucleon interactions down to $2 \cdot 10^{-45}$ cm^2 cross-section at a WIMP mass of 100 GeV/c^2 by the end of 2009; and the upgraded XENON-100, with a low-background copper vessel, new QUPID photo-sensors and better krypton separation, should improve this sensitivity by an order of magnitude by 2012 (Aprile and Baudis, 2008).

The muon flux at the 3100 mwe LNGS depth is about 22 m^{-2}d^{-1}. Neutrons generated in the rock walls and the passive shield of the detector is the dominant NR background. Adding 20 cm of polyethylene outside the current Pb shield reduce the neutron flux from the rock walls by a factor 10. Neutrons generated in the shield by cosmic muons yield a total NR background of 2.7±0.7 single NR events per year in a 100 kg target. To reduce the background, a 98% efficient active muon veto arranged with plastic scintillators is used for the upgraded XENON-100 experiment.

The gas used in XENON-10 was purified from krypton at CWRU using a *chromatographic column* to achieve < 3ppt* concentration of krypton (Bolozdynya, Brusov *et al.*, 2007). Krypton admixture is dangerous because it contains ^{85}Kr radioactive isotope (687 keV

---

* 1 ppt (part per trillion) = 10^{-12} relative concentration maximum energy beta-decay, 10.76 year half-life) at the level of ~ 1 MBq/kg or ~ 1 ppt relative concentration in a natural mixture of krypton isotopes (Anisimov *et al.*, 1989b).

The gas used in XENON-100 was processed by the Spectra Gases Company using their distillation plant. During the first background run with XENON-100, krypton was found to be present in the xenon at a level of < 50 ppt. The updated XENON-100 will use xenon purified with a custom cryogenic distillation column produced by Taiyo-Nippon Sanso Company. This purification system is designed to reduce the krypton concentration by a factor of 1000 in one pass at a purification speed of 0.6 kg/h.

UK Dark Matter Search

The ZEPLIN II detector, with a 31 kg LXe active mass as described in Chapter 6 (Fig. 6.14), was commissioned for a WIMP search 1100 m underground at Boulby Mine (England) in the fall of 2005. Shielding included a liquid scintillator veto (30 cm radial) and Gd-loaded HDPE (top/bottom) and 25 cm of Pb (all directions). The detector used 5" ETL D742Q PMTs mounted above the LXe surface (Alner *et al.*, 2007).

A WIMP search was conducted during the spring of 2006, and data was collected for 31 days; this included daily, automated energy calibrations. A gamma calibration at 1.0 kV/cm drift field yielded 0.55 phe/keVee, in agreement with Monte Carlo simulations. The xenon purification system with continuous gas circulation achieved an electron lifetime in excess of 1 ms (2 m electron drift path). A blind analysis procedure was defined, in which a 10% subset of the data was released immediately for analysis. A population of low S2/S1 events was seen at a large radius (i.e. near the chamber wall). Such events have also been observed by other groups, and they are possibly caused by the decay of radon daughter nuclei deposited on the inside of the PTFE wall, for which an alpha decay into the wall causes recoil of the heavy nucleus into the liquid. These events are vetoed by radial cuts; however, since the PMTs are large (∅12.5 cm), the position reconstruction is accurate only to about 5 cm, this makes it necessary to cut 50% of the radius (leaving 8 kg fiducial). Cuts are also made on multiple scatters, on the time interval between S1 and S2 (to remove alpha-particle related events close to the top grid and cathode).

From the observed and expected event count, a 90% confidence upper limit, obtained by the Feldman–Cousins method, to the number of nuclear recoil events was observed within the defined acceptance window. For the combined energy span of 5–20 keVee, where 29 events are seen and 28.6±4.3 are expected, this yields a 90% CL of 10.4 nuclear recoil events within the 50% nuclear recoil acceptance window in 225 kg·d of exposure, using the mean expectation value, or an upper limit of 0.092 events/kg/day in total between 5 and 20 keVee. This converts to an upper limit for the WIMP–nucleon spin-independent cross-section with a minimum of $6.6 \cdot 10^{-43}$ cm^2 (Alner *et al.*, 2007).

The next detector of the ZEPLIN detector family — *ZEPLIN-III* — uses 12 kg of LXe viewed by 31 2"-diameter ETL D730/9829Q photo-multipliers immersed in the liquid and located under the grid cathode (Akimov *et al.*, 2007). The electric field in the target volume is defined by a cathode wire grid 36 mm below the liquid surface and a reflective anode plate 4 mm above the surface in the gas phase. During normal operation, the electric field strength is 3.9 kV/cm in the liquid, and 7.8 kV/cm in the gas. A fiducial volume containing 6.5 kg LXe was defined by using a time window between S1 and S2 signals and by two-dimensional position reconstruction from the PMT signals in the central area of 300 mm diameter and 3.5 cm depth.

Similar to the first electroluminescent emission detectors (see Chapter 5) ZEPLIN-III has an electrode system without separate grids arranging electron drift and electron extraction from the liquid and has no Teflon reflector. As a result, the detector was operated at high drift electric field and provided effective light collection at the cost of a relatively shallow (3.5 cm) fiducial volume that did not allow taking advantage of the self-shielding effect from radioactive background that originated from the photo-multipliers. Nevertheless, after a fiducial exposure of 450 kg·d at the Boulby mine, the ZEPLIN-III provided the best upper limit on the WIMP-nucleon spin-dependent cross-section up to that time: $1.8 \cdot 10^{-2}$ pb for a 55 GeV/c^2 WIMP mass at a 90% confidence level (Lebedenko *et al.*, 2009).

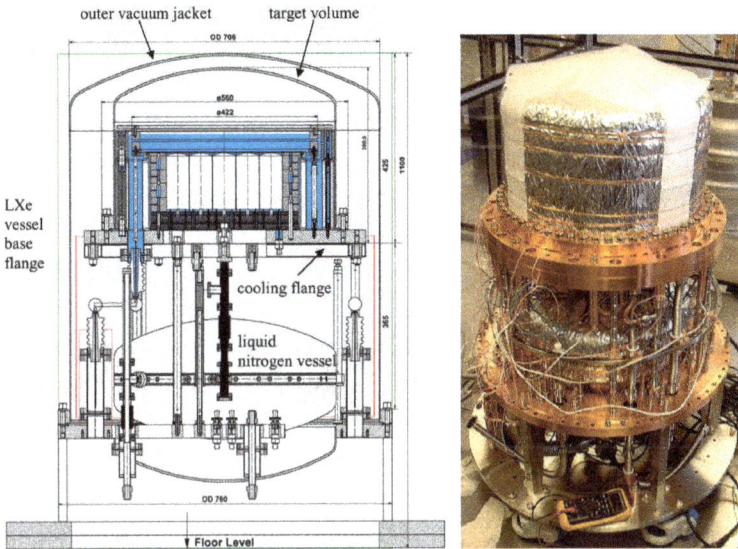

Figure 8.3 ZEPLIN-III detector: a schematic drawing in cross-section view (left) and a picture of the instrument with removed the outer vacuum vessel (right). Courtesy of T. Sumner.

LUX experiment

The goal of the LUX experiment is to clearly detect or to exclude WIMPs with a spin independent cross-section per nucleon of $7 \cdot 10^{-46}$ cm^2, equivalent to ~ 0.5 event/100 kg/month in an inner 100 kg fiducial volume (FV) of a 300 kg LXe detector (Fig. 7.6). The overall background goals are to ensure < 1 background events characterised as possible WIMPs in the FV in ten months live time. The backgrounds consist of electron recoils, primarily from gamma rays, and nuclear recoils from neutrons. The combined gamma and beta event rate goals in LUX are $< 8 \cdot 10^{-4}$ events/keVee/kg/day at a threshold energy within the detector's FV.

The three-dimensional imaging capability of the detector with a bulk working medium is a powerful tool for eliminating a majority of the background events which occur in the outer portions of the detector, and will also eliminate resolvable multiple-scattering events. The effectiveness of these rejection methods depends strongly on the specific detector design. Extensive Monte Carlo simulations have been performed for the large LUX detector, and show that the necessary suppression can readily be achieved.

Most modern dark matter experiments, including CDMS, XENON and ZEPLIN, have utilised a shield, which combines Pb shielding for gamma rays, with a moderator, either polyethylene or active liquid scintillator, to attenuate or veto neutrons to the desired level. The LUX detector will be immersed in a $185\,\mathrm{m}^3$ water tank used to shield the detector from backgrounds radiated by the cavern rock (Fig. 8.4). With a minimum water shield thickness in any direction of 2.5 m, the cavern sources of gamma and neutron activities will be reduced to a level where the internal radioactivity from detector components is the dominant source of gamma and neutron background. Monte Carlo simulations show that a 2.5 m layer of water is able to reduce the residual flux of $> 100\,\mathrm{keV}$ neutrons entering the detector by more than a factor of 100.

Figure 8.4 Artistic view of the LUX detector installed into water tank shielding located at the Davis's cave of Homestake mine at 4.5 km water equivalent depth. Courtesy of the LUX collaboration.

A significant contribution to the ER event rate within the LXe is expected to come from the PMTs. The neutron recoil background event rate detected in the LXe can be suppressed by requiring that events occur as single scatters within the FV. For a FV of 100 kg, the event rate due to a neutron emission spectrum of the type typical for neutrons produced in (α,n)-reactions is typically suppressed by a factor > 20, compared to the average neutron event rate for the entire active volume.

An additional suppression factor 3 will be achieved by vetoing the neutron recoil event, since some part of the gamma cascade (> 8 MeV total energy) associated with thermal neutron capture in xenon and deposited in the working medium can be detected.

The most serious background problem comes from trace levels of krypton. The LUX goal is to reduce the total concentration of any krypton atoms in xenon below 4 ppt, using a chromatographic separation system specially developed for XENON-10 experiment (Bolozdynya *et al.*, 2007) and upgraded in capacity for LUX.

The other problematic radioactive noble gas impurity is noble gas Rn, which can emanate from a variety of sources inside the detector, and has been shown by several experiments to decay in the bulk fluid. To match the LUX ER background, Radon decay rate of ~ 16 mBq are required. In XENON-10, the observed bulk α-decay rate is 4 mBq. In ZEPLIN-II, a ~ 1 Bq rate was observed, but it was found to emanate entirely from the particular getter used at that time by ZEPLIN-II for Xe purification.

A highly penetrating neutron background arises from high energy cosmic ray muons interacting in the cavern rock. Neutrons, like WIMPs, deposit energy through nuclear recoils. Neutron scattering events at low energies must be reduced below the rate of any possible WIMP signal. However, neutrons — unlike WIMPs — will undergo multiple scattering within the active volume of a detector. In a detector able to resolve multiple vertices, this can be used to veto potential contributions from neutrons. The water shield is designed to ensure that all external neutron sources will be suppressed to a level that is well below the expected WIMP interaction rate, corresponding to < 10^{-46} cm^2 scalar cross-section, independent of any additional cuts made in the detector. This will aid the unambiguous identification of nuclear recoil signals caused by WIMPs.

To reach below 10^{-9} pb, a detector of 1–10 tonne mass is required (Spooner, 2003; Aprile *et al.*, 2005). The next generation of the LUX detectors family, currently under development, will achieve up to 20 tonnes active mass.

XMASS experiment

The *XMASS* project is a multipurpose ultra-low background experiment using liquid xenon as detector medium (Namba, 2005). Its goals are the following:

1) to observe low energy solar neutrinos (xenon massive detector for solar neutrinos);

2) to detect WIMP dark matter (xenon detector for weakly interacting massive particles);

3) to find the neutrino-less double-beta decay of ^{136}Xe (neutrino mass detector).

Under this project three detectors have been constructed and tested: a 3 kg LXe emission detector as described in Chapter 6 (Fig. 6.12), a 100 kg R&D prototype scintillation detector, and a 1 m diameter 'wall-less' scintillation LXe detector with about 1 tonne fiducial volume mass for underground experiments.

In order to reduce the internal background, a gas distillation system was developed, which allows reduction of the krypton concentration to < 4 ppt (Takeuchi, 2004).

The emission detector with ~ 300 cm^3 FV was installed in the Kamioka mine in 2001 and used for background study. The physics result was one order of magnitude worse than the best limit at that time for the WIMP-nucleon cross-section of spin-independent couplings (Yamashita, 2003).

The 100 kg option for the XMASS detector is a cubic copper chamber with the inner sides having a length of 31 cm, each being viewed with nine Hamamatsu R8778 PMTs. In order to reduce the background, the experimental detector was placed in the Kamioka Mine to a 2700 mwe depth and enclosed within a passive multi-layer shield with an outside dimension of ~ 2 m. With the large detector, the experiment is expected to achieve a $3.3 \cdot 10^{26}$ year half-life time limit for

$0\nu\beta\beta$ (^{136}Xe) decay and 30 DM events per day for 100 GeV 10^{-6} pb SI for proton; 10 pp-neutrino and 5 ^{7}Be-neutrinos per day.

WARP experiment

The WARP (WIMP Argon Program) is an experimental program aimed at the detection of WIMP elastic interactions occurring in a liquid argon target. The identification of nuclear recoils and suppression of the gamma-ray background is provided by the following two measures:

1) the analysis of the decay time of the scintillation (S1 signal) in liquid argon;

2) the comparison of the scintillation (S1) and electroluminescent signal (S2) occurring during collection of ionization electrons in the gas phase.

The liquid argon scintillation contains two components with two basic decay times, which are very different (see, for example, Hitachi *et al.*, 1983): $\tau_s = 7.0\pm1.0$ ns and $\tau_t = 1600\pm100$ ns — and their relative intensity is sensitive to the density of ionization or to the nature of the ionizing particles. A ratio of the integral intensities I_s / I_t is found to be 0.3, 1.3 and 3 for electron, α-particle and ion recoil excitations in LAr, respectively. Depending on the values of the applied electric field, the ratio of the scintillation and electroluminescence signals S is about 180 for minimum ionizing particles, 3 for α-particles and 10 for Ar-ion recoils.

If employed, the two independent signal analyses provide effective identification of the recoils. Recently (Benetti *et al.* 2008), the collaboration reported a limit close to the result of XENON-10 achieved with 96.5 kg·day exposition of the test WARP-10 detector containing 2.6 kg LAr in the sensitive volume (Fig. 6.15) at the Gran Sasso Underground Lab.

The next goal of the program is the construction of a 100 L sensitive volume LAr emission detector. Argon is less sensitive than xenon to electro-magnetic radiation, which composes the majority of the natural background. However, there is a dangerous β-active isotope ^{39}Ar, which in natural mixture of isotopes has an activity of about 0.76 Bq/kg. The depletion of the ^{39}Ar isotope would require a very expensive process.

The authors of the WARP program expect to effectively utilise rejection factors that can suppress the background from beta-decay by a factor of the order of $10^{-7}–10^{-8}$. Then, the surviving background count rate from ^{39}Ar can be reduced to the level of one event in 10–100 days — low enough to improve the existing limit on WIMP cross-section by two orders of magnitude. Recently, Acosta-Kanea *et al.* (2008) reported that argon from underground natural gas reservoirs contains a low level of ^{39}Ar. In the gas stored in the US National Helium Reserve, the ratio of ^{39}Ar to stable argon was measured to be $\leq 4 \cdot 10^{-17}$ (84% CL), less than 5% of the value in atmospheric argon (^{39}Ar/Ar $= 8 \cdot 10^{-16}$). The total quantity of argon currently stored in the National Helium Reserve is estimated at 1000 tonnes.

Using argon opens the prospect of constructing massive detectors due to its high abundance in nature. Since the ICARUS collaboration demonstrated the feasibility of a multi-tonne LAr TPC, kilo-ton emission detectors are now being proposed for the detection of atmospheric, solar and supernova neutrinos and to search for proton decay (Rubbia, 2003; Ereditato and Rubbia, 2005, 2006).

With the increasing sensitivity of dark matter experiments searching for direct interactions (Table 8.1), the experimental technique is approaching the point when neutrino interactions will become an

Table 8.1 WIMP emission detectors: completed, running or under construction in 2009.

Project	Detector mass, Total/Feducial, kg	Sensitivity, 10^{-44}cm^2	Location, Years on duty	Status	Ref.
XENON-10	25/5 LXe	8.8 @ 100 GeV/c^2 5.5 @ 30 GeV/c^2	GS, 2006-07	Completed	a
XENON-100	100/10 LXe	0.2 @ 100 GeV/c^2	GS, 2008-09	Active	b
ZEPLIN II	31/8 LXe	66 @ 55 GeV/c^2	BM, 2006-07	Completed	c
ZEPLIN III	12/6.5 LXe	0.18 @ 55 GeV/c^2	BM, 2008-09	Active	d
LUX	300/100 LXe	0.07 @ 100 GeV/c^2	H, 2010	u/c	e
WARP-10	10/2.6 LAr	75 @ 100 GeV/c^2	GS, 2006	Completed	g
WARP-100	100 LAr	1 @ 100 GeV/c^2	GS, 2009-10	u/c	g

Notes: (BM) Boulby mine (England); (GS) Gran Sasso Underground Laboratory (Italy); (H) Homestake DUSEL (South Dakota); (u/c) under construction; (a) Angle *et al.*, 2007; (b) Aprile and Baudis, 2009; (c) Alner *et al.*, 2007; (d) Lebedenko *et al.*, 2009; (e) courtesy of LUX collaboration, 2009; (g) Benetti *et al.*, 2008.

irreducible source of background of experiments searching for WIMPs and can be considered as other objects of interest. In Fig. 8.5 there is presented the data from Arisaka confirming this statement and also demonstrating that WIMP detectors of the next generation with multi-tonne masses of working media even with natural mixture of isotopes become sensitive to double-beta decay and solar neutrinos and can be used in multi-task experiments.

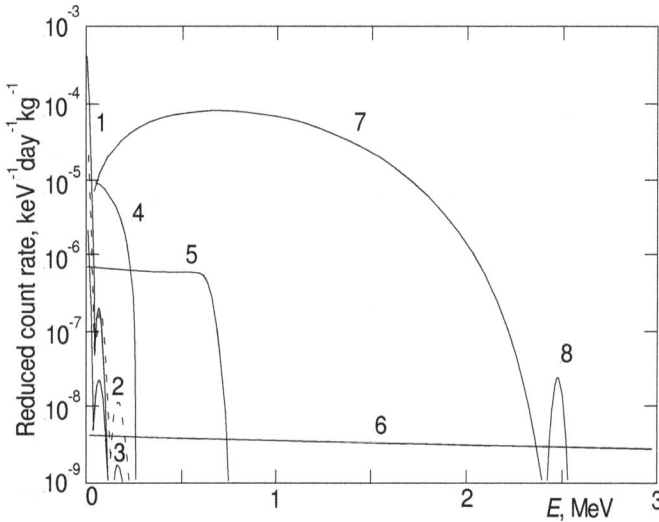

Figure 8.5 Expected energy spectra of WIMPs of 0.1, 1 and 10 TeV/c^2 mass (1, 2 and 3, respectively), solar neutrinos of pp, ^{7}Be and ^{8}B cycles (4, 5 and 6, respectively) and double beta decays with 10^{22} for neutrino (7) and 10^{27} years life times for neutrino-less (8) decays, respectively, in natural mixture of xenon isotopes used as a target material. Redrawn from Arisaka *et al.*, 2009.

8.2 Neutrino detectors

Neutrino interactions of interest include (ve)-neutral current elastic scattering, where the neutrino interacts with the atomic electrons, and (vA)-neutral current coherent elastic scattering, where the neutrino interacts with the target nucleus. The former process has been considered as a method for solar neutrino detection in low-threshold detectors with cross-sections of order 10^{-44} cm^2 and maximum recoil electron kinetic energy up to a few hundred keV. The latter process can be used for the detection of reactor neutrinos but it has never been observed in spite of having a relatively large cross-section of order 10^{-39} cm^2 because the maximum nuclear recoil kinetic energy is very low and expected to be in the range of a few keV (Monroe and Fisher, 2007).

Detection of low-energy neutrinos was considered as one of the first prospective physical tasks for noble gas emission detectors (Rodionov, 1979; Bolozdynya *et al.*, 1995).

8.2.1 Measurement of magnetic moment

One fundamental but never measured property of the neutrino is the *magnetic moment*. The magnetic moment may help to explain the solar neutrino deficit and the anti-correlation of the measured neutrino flux with solar magnetic activity. The presence of a magnetic moment can essentially influence the rate of neutrino interactions in nuclear reactors and stars, and affect many theoretical predictions in the evolution of the universe. The lowest limits on the neutrino magnetic moment come from both solar neutrino experiments and reactor experiments. Data from the Super-K experiment set the tightest limit for solar neutrino to date $\mu_V < 1.5{\cdot}10^{-10}\,\mu_B$ at 90%CL (Beacom and Vogel, 1999) and the reactor experiment GEMMA (Beda *et al.*, 2009) recently achieved the best limit $\mu_V < 3.2{\cdot}10^{-11}\,\mu_B$.

Since 1992 a liquid xenon detector has been considered to probe anti-neutrino/electron scattering cross-sections in the range of < 100 keV deposited energy where the cross-section of elastic $\overline{V}_e e$-scattering of a neutrino with a magnetic moment may exceed the electroweak cross-section due to additional magnetic dipole-dipole scattering (Cline and Hong, 1992). Such an experiment could be performed with a moderate size (~ 1 t mass) of LXe emission detector (Baldo-Ceolin *et al.*, 1992), with an artificial neutrino source, and it could achieve a sensitivity of < $3{\cdot}10^{-12}\,\mu_B$ (Barabanov *et al.*, 1997).

The LXe emission detector for the neutrino magnetic moment measurement was designed and constructed at ITEP (Moscow) in 1994–1998 as an emission 'wall-less' camera with a single array of seven 2" diameter photo-multipliers placed in the gas phase above the liquid xenon. The array was viewing ~ 1 cm gap between two grids. The top (anode) grid was placed in the gas phase, the bottom one was placed just below the liquid surface. The electrode system included the anode and the bottom grid, drift ring electrodes and a thin aluminium cathode placed in the liquid and, as an option, coated with wavelength shifter. The liquid was contained in a 30 litre vessel made of low-background titanium. A schematic drawing of the detector and a photograph of the electrode structure at the moment of installation inside the titanium vessel is shown in Fig. 8.6. For purification of xenon a large (100 L capacity) spark purification system has been developed (Anisimov *et al.*, 1991) and has demonstrated a capability to provide > 100 cm electron drift in multi-kilogram samples of LXe and LKr.

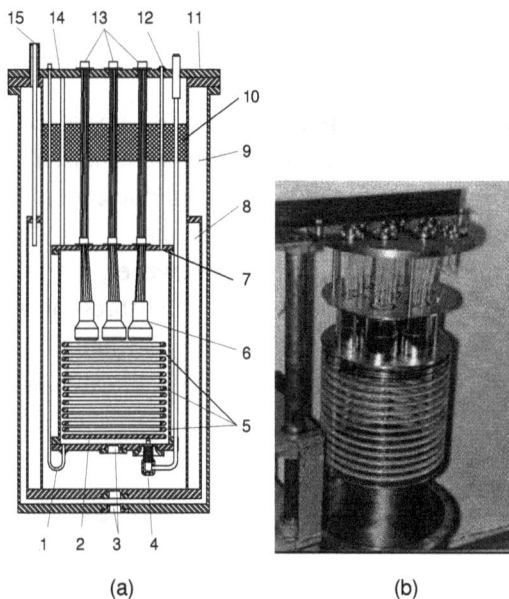

(a)   (b)

Figure 8.6 LXe emission detector for neutrino magnetic momentum measurement constructed at ITEP (Moscow) in 1994–1998: a schematic drawing (a) and a picture of installation of the insert structure into the low-background titanium cryostat (b); LXe inlet (1); cathode (2); X-ray input windows for purity measurements (3); high voltage feedthrough (4); drift field electrodes (5); photo-multiplier (6); gas outlet (7); liquid nitrogen jacket (8); vacuum cryostat (9); foam thermoinsulation (10); flange (11); gas connector (12); photo-multiplier feedthrough (13); rod supporting cold titanium vessel (14); liquid nitrogen inlet/outlet (15).

Unfortunately, this project has never been finished because of insufficient funding during the period of transformation of USSR to Russian Federation in 1990s of the past century.

8.2.2 Measurement of coherent scattering

Direct detection experiments searching for WIMPs as described above in Section 8.1.2 use the *coherent elastic scattering* process. Ordinary neutrinos of energy 1–10 MeV also can interact coherently with atomic nuclei causing the nuclei to recoil with energies up to tens of keV. The cross-section of such a process depends on the neutron number N and the neutrino energy E_V as follows (Drukier and Stodolsky, 1984)

$$\sigma \approx 0.4 \cdot 10^{-44} N^2 (E_v)^2 \ \text{cm}^2 \tag{8.1}$$

where E_v is measured in MeV.

This formula is valid for neutrino energies up to about 50 MeV, and thus applies to reactor, solar and supernova neutrinos. The magnification factor N^2 gives a significant count rate from nuclear reactors and natural sources in relatively small detectors that may have a great impact on nuclear reactor monitoring techniques. However, a signal consisting of only a few ionization electrons is below the threshold of conventional solid or liquid state detectors without internal amplification. For fixed neutrino energy, the recoil spectrum falls linearly with the average energy of

$$\langle E_r \rangle = \frac{1}{3} E_r^{\text{max}} = 716 \left(\frac{(E_v)^2}{A} \right) \text{eV} \tag{8.2}$$

where E_v is measured in MeV and A is the atomic number of the target nucleus.

There is proposed a very compact (kg-scale) liquid argon emission detector to measure coherent neutrino scattering (Hagmann and Bernstein, 2004) as shown in Fig. 8.7. Argon has been chosen as a working medium because of its more favourable kinematics of nuclear recoils compared to xenon and less sensitivity to gamma background.

For the neutrino flux at 25 m distance from a 3 GW reactor, the expected event rates before detection efficiencies are 56 $kg^{-1}day^{-1}$ for coherent scattering off argon compared to 2.8 $kg^{-1} day^{-1}$ for the inverse beta decay reaction in $(CH)_n$ usually used for detection of reactor neutrino scatters. This allows the construction of a very compact and potentially mobile detector of low-energy neutrinos.

Results of computer simulations of the expected effect in comparison with gamma- and neutron-induced backgrounds are shown in Fig. 8.8. It was suggested that the detector be surrounded by 20 cm low-background concrete. A reduction of the external background a factor ~ 100 is achieved with a shield consisting of 2 cm lead (inner) and 10 cm of borated polyethylene (outer). The strongest background is expected from the beta decay of ^{39}Ar ($E_{max} = 565$ keV, $\tau_{1/2} = 269$ y), which induces a total radioactivity of ~ 1 Bq/kg in natural argon and an estimated differential beta activity of ~ 4 mBq/keV/kg at the low energy end of the spectrum. To reduce this background, isotopic separation of the working medium is needed.

8.2.3 Detection of neutrinos from the sun

The detection of low-energy neutrinos from the sun is one of the most challenging tasks. Most successful experiments, relying on radio-isotope analyses (R. Devis experiment at Homestake, GALLEX/GNO and SAGE) in chlorine and gallium and on production of Cerenkov radiation in water (Kamiokande, SuperKamiokande and SNO), are sensitive to the upper range of the solar neutrino energy spectrum (~ 1–10 MeV). However, this window contains only a small fraction of the total solar neutrino flux, about 98% of which is expected to have energies below 1 MeV (YongLin *et al.*, 2007) and only Cerenkov detectors provide information about the energies of the neutrinos. The next major step in exploration of solar neutrino emission is to measure neutrino fluxes in the low energy range and, in particularly, to measure the flux from the *pp* reaction. These neutrinos have the energy spectrum continued up to 420 keV, with peaks in the range of 200–300 keV, and their flux on the earth is precisely predicted ($5.94 \cdot 10^{10}$ $s^{-1}cm^{-2}$) by the standard solar model. Checking this prediction would be groundbreaking.

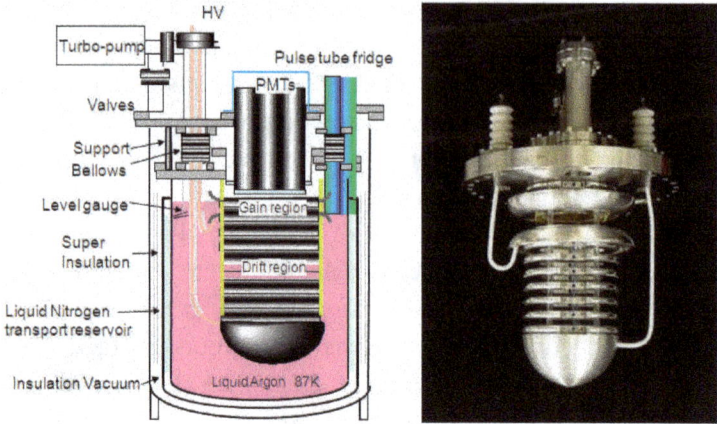

Figure 8.7 Liquid argon coherent scattering neutrino detector schematic view (left) and picture of the electrode system (right). Courtesy of A. Bernstein.

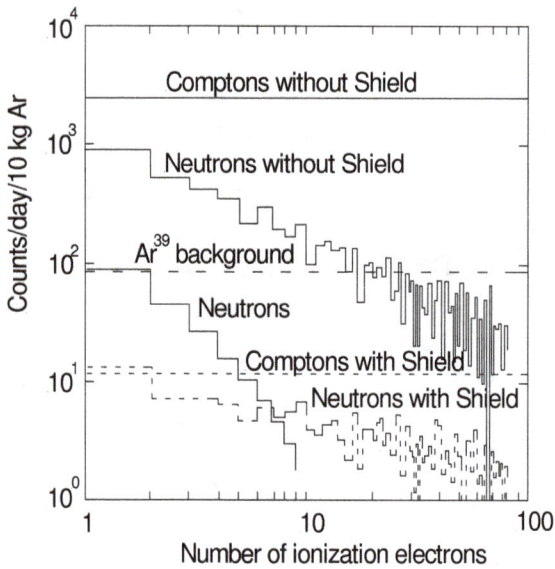

Figure 8.8 Monte Carlo simulations of the effect of coherent neutrino scattering in comparison with gamma and neutron induced backgrounds. Redrawn from Hagmann and Bernstein, 2004.

HERON project

The *HERON project* is an effort to develop a detector for low-energy solar neutrinos produced in the *pp* and ^{7}Be solar fusion reactions by observing their elastic scattering

$$V_x + e \rightarrow V_x + e \qquad (8.3)$$

using superfluid helium as a target (Sethumadhavan *et al.*, 2004). The elastic cross-section is very precisely known and requires no external artificial sources for calibration. In addition, the reaction is sensitive to all neutrino flavours and can improve knowledge of the solar neutrino flux and the mixing angle between the so-called m_1 and m_2 mass eigenstates.

In the proposed detector with a fiducial volume of 70 m^3, 20 events per day are to be expected according to the Standard Solar Model. In neutrino elastic scattering, the recoil electron deposits its energy in liquid helium in a few ways: ionizing atoms and generating photons and rotons in the liquid ^{4}He superfluid state (density 0.145 g/cm^3). Scintillation and ionization signals will be used for detection purposes. The detector design is similar to the LXe/LAr emission detectors discussed in Chapters 6 and 7 but both electrons and photons will be detected with an array of superconducting bolometers installed above the liquid that allows essentially to reduce the detection threshold and even to use a coded aperture for determination of the direction to the photon source (Huang *et al.*, 2008).

Since helium has a low stopping power for gamma radiation, self-shielding is not so effective as in liquid xenon detectors, so the detector will be additionally shielded against gamma rays with solid nitrogen (104 tonne mass, 1.03 g/cm^3 density) enclosed in acrylic cells (5.6 tonne mass).

Additional rejection of gamma backgrounds can be achieved using analyses of the spatial distribution of ionization electrons. The maximum track length expected for a neutrino event is ~ 2cm, which effectively looks like a point source of the scintillation light. The gamma background events look like multiple depositions over an average length of more than 50 cm in liquid helium. Consequently, electrons distributions for signal and background will be different. These features will be used for rejection of the gamma background.

Superfluid helium has the unique property of 'self-cleaning'. At the operating temperature ~ 30 mK of the superfluid, thermal energy is sufficiently low relative to the gravitational potential energy that other atomic species falls out of the bulk. The detector is supposed to be installed in a 3.5 m thick water shield at Homestake DUSEL.

Simulations predict that HERON will be capable to measure *pp* and ^{7}Be fluxes with ±1.68% and ±2.97% accuracy, respectively, in a five-year data sample (Huang *et al.*, 2008).

XAX project

A multi-tonne emission detector system XAX, comprised of concentric 10-tonne liquid targets of ^{136}Xe and $^{129/131}$Xe together with similar liquid argon target, is proposed to segregate signals from WIMPs, solar neutrino and double-beta decay (Arisaka *et al.*, 2009). The individual detector of this system is shown in Fig. 7.7. The use of multiple target elements allows for confirmation of A^2 dependence of a coherent scattering cross-section, the different isotopes provide information on the spin-dependence of the dark matter interaction and detection of neutrinoless double-beta decay with lifetimes of 10^{27}–10^{28} years, corresponding to the Majorana neutrino mass range 0.01–0.1 eV. The use of a ^{136}Xe depleted LXe target will allow the measurement of the *pp* solar neutrino spectrum and flux with an accuracy of 1–2%. The system is supposed to be installed at Homestake DUSEL.

8.3 Double beta-decay detectors

The neutrino oscillation experiments indicate that neutrinos do have mass and, in particular at least one neutrino has a mass greater than 50 meV. The upcoming $\beta\beta(0\nu)$ experiments with large noble gas detectors will have good enough sensitivity to check below this critical mass scale. It is expected that a positive result will be achieved within the coming decade (Elliot and Vogel, 2002).

8.3.1 Positron double-beta decay

Searching for neutrino-less transitions in positron decays gives a unique signature because of the possibility to select these extremely rare events in coincidence with annihilation photons (Staudt *et al.* 1991). The clear signature is competing with very long lifetimes predicted for double-positron decays, which are in several orders of magnitude larger then those for double-electron decays (Zeldovich and Khlopov, 1981). Two of seven famous isotopes, which are able to decay in this way, are noble gas isotopes of ^{124}Xe and ^{78}Kr.

A new experiment was proposed (Bolozdynya, Egorov *et al.*, 1997) to search for *positron double-beta decays* of ^{124}Xe and ^{78}Kr using the three-dimensional position-sensitive emission detector of ~ 1 m^3 total volume filled with liquid xenon, liquid krypton or their mixture and optionally surrounded with scintillators. The emission detector is triggered by scintillations occurred in the liquid at the moment of the decays happened.

The unique feature of the proposed experiment is the very specific topology of the useful events that make the experiment comparable in sensitivity with tracking experiments (see, for example, Artemiev *et al.*, 2005 and references therein). In $(2\beta^+0\nu)$-decay case, the useful event will contain five point-like ionization clusters (vertices), cross-laying in the same plane (Fig. 8.9). The central vertex is a short track of two positrons with 822 keV total deposited energy in the case of ^{124}Xe $(2\beta^+0\nu)$-decay and with 833 keV total deposited energy in the case of ^{78}Kr $(2\beta^+0\nu)$-decay. Four surrounding vertices are placed in the circle of about 15 cm radius with ~ 80% probability, and these are point-like tracks of Compton electrons occurring at first scatters (or photo-absorptions with smaller probability) of 511 keV annihilation photons. In the case of the $K\beta^+(0\nu)$-capture, the useful event will contain three vertices, located along the straight line. The central vertex will have 1844 keV(^{124}Xe) or 1855 keV (^{78}Kr) deposited energy, and two Compton vertices are placed at both sides in the range of 15 cm with 90% probability. Note that the specific sizes of flat $2\beta^+$ 'Compton Cross' and $K\beta^+$ 'Compton Line' are practically the same in the liquid xenon and in the liquid krypton at their triple points.

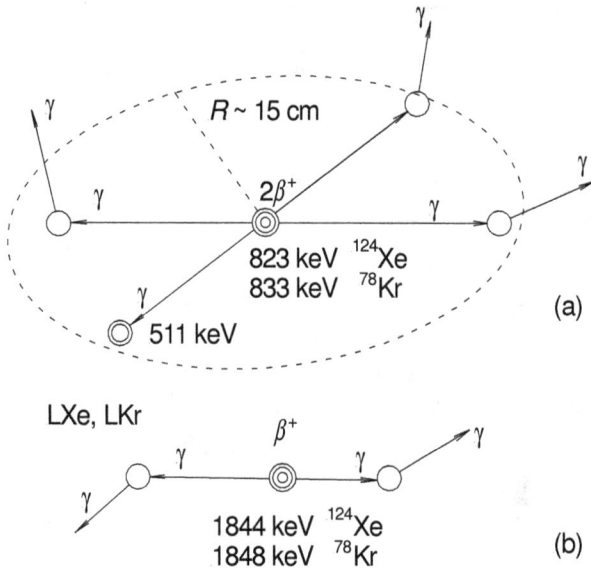

Figure 8.9 Images of neutrinoless double positron decay (a) and K-capture (b) of ^{78}Kr and ^{124}Xe in liquid krypton or liquid xenon. Redrawn from Bolozdynya, 2006.

In the cases of $(2\beta^+2\nu)$-decay and $K\beta^+(2\nu)$-capture, the pictures look the same but the energy deposited in the central vertex is less certain.

The imitation of such specific events by natural backgrounds has negligible probability, even if the detector is located in the conditions of a normal above-ground laboratory. A large mass of the working medium makes it feasible to use a natural mixture of isotopes and can still achieve a very high sensitivity. Note that the intense background from beta-decay of ^{85}Kr in the natural mixture of krypton isotopes requires triggering in coincidence with two or four outside signals from scintillators surrounding the detector and absorbing Compton scattered gamma rays. Exposing the detector filled with the natural mixtures of isotopes in the underground laboratory makes it possible to search additionally for $2\beta^-$ decays of ^{134}Xe (~ 50 kg content inside the fiducial volume enclosed inside 15 cm thick LXe internal shielding) and for ^{136}Xe (~ 30 kg content).

8.3.2 Electron double-beta decay

The Enriched Xenon Observatory (EXO) experiment is a search for $0\nu\beta\beta(^{136}\text{Xe})$ decay with a massive (up to 10 tonne) xenon detector including a system for laser tagging ^{136}Ba daughter nucleus (Danilov *et al.* 2000). The double-beta decay of ^{136}Xe produces a Ba^{++} ion in the final state that can be neutralised to Ba^+ in the TPC. The single barium ions can be effectively detected with fluorescence induced by the intensive laser excitation (Moe, 1991). Ground-state ions can be optically excited to the $6^2\text{P}_{1/2}$ state from where they have a 20% branching ratio to decay into the meta-stable $5^4\text{D}_{1/2}$ state, then specific Ba^+ detection may be achieved by exciting the ion from the meta-stable state back to the $6^2\text{P}_{1/2}$ state and observing the 493 nm blue photon of transition back to the ground state with branching ratio 70%. This procedure may be repeated many times in order to produce a significant signal.

The EXO collaboration is currently focussed on the high-pressure gas TPC technology but at the stage of R&D considered a liquid xenon detector with extraction of Ba^+ ions from the liquid as a promising option. The TPC baseline design consists of two 35 m^3 modules filled with 8.4 t xenon at about 20 bar pressure. The gas would be contained in a non-structural bag within a pressurised buffer gas. The spatial resolution will permit the identification of the high-ionization-density points at the terminus of the beta tracks, aiding in the separation of two-electron events from one-electron backgrounds such as Compton scatters. Being triggered by a PMT array, detecting a well-defined deposition energy 2.481 MeV as a scintillation flash, the lasers are directed to the decay point to excite the Ba^+ ion. One complication is that the originally produced Ba^{++} ion needs charge exchange with the detector medium but Xe is a tightly bound atom and this process is unlikely in pure xenon. Adding a quenching gas may be used if scintillation light yield will not be suppressed.

The advantage of the liquid xenon version of the detector is its inherent compactness along with the self-shielding property acts effectively against gamma backgrounds. However, this approach requires the development of measures on extraction of ions at extremely high electric fields (see Section 2.4.3) that are difficult to apply to the gas phase without breakdown. One of the possible solutions of this problem is to add a pressurised non-condensing ballast gas such as helium to the gas phase of the emission detector.

8.4 Imaging radiation fields

Since the discovery of electron conductivity in noble liquids, there is an interest in using liquid xenon detectors for imaging radiation fields, for example, in nuclear medicine. Condensed Xe provides high stopping power for electrons similar to NaI(Tl), successfully used for gamma imaging for 50 years, and at the same time, it is available in a large amount and can be used in detectors with very large fields of view. Emission detectors with internal amplification of signals may extend the gamma imaging technique in the range of small energies and three-dimensional imaging in Compton camera mode.

8.4.1 Emission camera for SPECT

In medicine, ionization radiation is used for imaging and diagnostics of functioning inner organs. Development of any new radiation detection technique is traditionally passing attempts to apply it to medicine needs. Very few of them are successful but all of them cultivate the ground for revolutionary developments in the future. The idea of using the unique combination of detecting properties of liquid xenon for medical imaging, stimulated by the significant technological progress, continues to interest physicists. With respect to γ-ray detection, LXe is comparable to NaI(Tl): it has a similar atomic number and density. Its scintillation light yield is also similar to NaI(Tl) but the decay time of LXe scintillation is much shorter (for details, see Chapter 4). Another very important feature of liquid xenon, which none of the inorganic scintillators has, is a free behaviour of electrons liberated in the ionization process that can be easily extracted from the particle track by electric field and collected onto electrodes. The number of collected electrons produced in the liquid by gamma rays, usually used in medicine, is not very large but detectable (for instance, 511- keV photon produces about 30,000 electrons; 80% of them escape recombination in a field of ≥ 1 kV/cm). Detection of the ionization signal allows precision position measurements, while the scintillation can ensure the fast timing (Chepel, 1993).

In Section 6.2.1 we discussed the design and properties of the electroluminescence emission gamma camera (EEGC) shown in Figs. 6.7 and 6.8. The detector has been developed for two-dimensional gamma ray imaging in nuclear medicine at the beginning of the 1980s (Bolozdynya *et al.*, 1981; Egorov *et al.*, 1983; Bolozdynya *et al.*, 1985). The EEGC electroluminescence output signals from the array of 19 photo-multipliers have been processed similar to scintillation output signals from the Anger camera mode (for details see, for example, Kalashnikov, 1985). A storage oscilloscope of Tektronix 603 has been used to collect data and to display images of an alpha-source placed on the cathode and a lead multi-hole collimator illuminated with gamma radiation of ^{241}Am or ^{57}Co gamma sources placed outside the detector. The images have been captured with a camera from the screen of the oscilloscope and processed with a pulse height analyser to define position and energy resolution. The internal position resolution was measured to be 2.5 mm FWHM, 3.5 mm FWHM position resolution was measured with a 59.6 keV (^{241}Am) gamma source located outside, and an energy resolution of 16% FWHM was measured with a 122 keV (^{57}Co) gamma source. These results improve upon the previous records achieved in similar size detectors such as NaI(Tl)-based Anger gamma cameras used for *single photon emission tomography* (SPECT).

The achievement was recognised by the USSR Academy of Science, which awarded the three youngest scientists working on this project with a medal of the USSR Academy of Science in 1983. However, at that time the new technique appeared to be too complicated and expensive for practical Soviet medicine. Nevertheless, from a technical point of view this development served as a predecessor of wall-less LXe emission detectors successfully used in upcoming experiments searching for cold dark matter as described above in Section 8.1.

8.4.2 Compton camera

Three-dimensional position sensitivity of EEGC detecting both scintillation and electroluminescence signals may serve as a basis for development of a LXe *Compton camera* for SPECT or industrial imaging without using physical lead collimators, which reduce the sensitivity of the detector (or acquisition time for radiation source with limited luminosity) by several orders of magnitude (see, for example, the review of 'Compton imaging techniques for nuclear medical imaging' by Rogers *et al.*, 2004). The accuracy of the Compton imaging depends on the uncertainties in the measurement the both the energies and three-dimensional localisation of a pair of Compton scatter and photo-absorption points of γ-ray interactions with the working medium. Since the xenon TPC can provide the accurate measurements of multi-vertex events and high detection efficiency for γ-rays, a prototype of a scintillation and electroluminescence Compton camera filled with pressurised xenon has been constructed and tested as a part of R&D efforts on development of the Compton camera for SPECT at the Nuclear Medicine Group of Siemens Medical Systems in the middle of the 1990s (Bolozdynya *et al.*, 1996). That was the first detector utilising both scintillation and electroluminescence of xenon to provide truly three-dimensional imaging of multi-vertex events with energy depositions in the range of 30–90 keV per vertex and 140 keV total depositions per event (Bolozdynya *et al.*, 1997). Soon after that it was understood that the Doppler broadening of the energy spectra of a Compton scatter electrons essentially limits the performance of the Compton camera based on heavy working medium at the energy range below 300 keV (Ordonez *et al.*, 1997; Rodgers *et al.*, 2004). Development of Compton cameras for high energy γ-rays (> 500 keV) requires a high-density working medium such as liquid xenon. This makes the idea of development of the emission Compton camera for imaging energetic γ-sources very attractive.

Features of LXe TPC Compton imaging for cosmic rays have been tested in the range of deposited energies of 0.4–4 MeV by Aprile *et al.* (2000) who have demonstrated the directional sensitivity of the TPC to γ-rays in Compton camera mode with angular resolution of 3° RMS at 1.8 MeV. The next step in the course of development of LXe Compton camera could be done using the emission detector technology.

8.4.3 Neutron imaging

Recently, a scintillation radiography technique has been used to map the distribution of ^{3}He impurity in liquid ^{4}He with a neutron beam, providing images of normal- and superfluid-component of ^{4}He velocity fields (Hayden *et al.*, 2004). Bearing in mind emission technology developed for the HERON project (Sections 2.3.4 and 8.2.3), one can suggest that enhancing the scintillation radiography with detection of ionization electrons extracted from the liquid helium containing ^{3}He by a two-dimensional array of bolometers allows to arrange *neutron imaging* similar to that considered above for gamma rays in emission detectors filled with heavy noble liquids. Hopefully, this approach will be investigated in the future.

Bibliography

Abramov A.V., Dolgoshein B.A., Kruglov A.A. and Rodionov B.U. (1975). Electrostatic emission of free electrons from solid xenon. *JETP Lett* 21: 82–85 (in Russian).

Acosta-Kanea D., Acciarri R., Amaize O., Antonello M., Baibussinov B., Baldo-Ceolin M., Ballentine C.J., Bansal R., Basgall L., Bazarko A., Benetti P., Benziger J., Burgers A., Calaprice F., Calligarichj E., Cambiaghi M., Canci N., Carbonara F., Cassidy M., Cavanna F., Centro S., Chavarria A., Cheng D., Cocco A.G., Collon P., Dalnoki-Veress F., de Haas E., Di Pompeo F., Fiorillo G., Fitch F., Gallo V., Galbiati C., Gaull M., Gazzana S., Grandi L., Goretti A., Highfill R., Highfill T., Hohman T., Ianni Al, Ianni An., LaCava A., Laubenstein M., Lee H.Y., Leung M., Loer B., Loosli H.H., Lyons B., Marks D., McCarty K., Meng G., Montanari C., Mukhopadhyay S., Nelson A., Palamara O., Pandola L., Pietropaolo F., Pivonka T., Pocar A., Purtschert R., Rappoldi A., Raselli G., Resnati F., Robertson D., Roncadelli M., Rossella M., Rubbia C., Ruderman J., Saldanha R., Schmitt C., Scott R., Segreto E., Shirley A., Szelc A.M., Tartaglia R., Tesileanu T., Ventura S., Vignoli C., Visnjic C., Vondrasek R. and Yushkov A. (2008). Discovery of underground argon with low level of radioactive [39]Ar and possible applications to WIMP dark matter detectors. *Nucl Instrum Meth A* 587: 46–51.

Adams J.S., Kim Y.H., Lanou R.E., Maris H.J. and Seidel G.M. (1998). Scintillation and quantum evaporation generated by single monoenergetic electrons stopped in superfluid helium. *J Low Temp Phys* 113: 1121–1128.

Afonasiev V.N., Akimov Yu. D., Bolozdynya A.I., Churakov D.L., Ermakov O.N., Kuzichev V.F., Lukashin V.M., Ovanesov M.V., Rogovsky I.A., Sushkov V.V. and Tchernyshev V.P. (1992). Detection of alpha-scintillations in condensed krypton by means of silicone photo-diode. *ITEP preprint 19–92*, ITEP: Moscow.

Ahmed Z., Akerib D.S., Arrenberg S., Attisha M.J., Bailey C.N., Baudis L., Bauer D.A., Beaty J., Brink P.L., Bruch T., Bunker R., Burke S., Cabrera B., Caldwell D.O., Cooley J., Cushman P., DeJongh F., Dragowsky M.R., Duong L., Emes J., Figueroa-Feliciano E., Filippini J., Fritts M., Gaitskell R.J., Golwala S.R., Grant D.R., Hall J., Hennings-Yeomans R., Hertel S., Holmgren D., Huber M.E., Mahapatra R., Mandic V., McCarthy K.A., Mirabolfathi N., Nelson H., Novak L., Ogburn R.W., Pyle M., Qiu X., Ramberg E., Rau W., Reisetter A., Saab T., Sadoulet B., Sander J., Schmitt R. Schnee R.W., Seitz D.N., Serfass B.A., Sirois A., Sundqvist K.M., Tarka M., Tomada A., Wang G., Yellin S., Yoo J. and Young

B.A. (2009). Search for Weakly Interacting Massive Particles with the First Five-Tower Data from the Cryogenic Dark Matter Search at the Soudan Underground Laboratory. *Phys Rev Lett* 102: 5.

Aihara H., Alston-Garnjost M., Badtke D.H., Bakken J.A., Barbaro-Galtieri A. *et al.* (1984). Search for $Q = 2/3e$ and $Q = 1/3e$ particles produced in e^+e^- annihilations, *Phys Rev Lett* 52: 2332–2336.

Akimov D. Yu., Afonasiev V.N., Bolozdynya A.I., Churakov D.I., Lamkov V.A., Sadovsky A.A., Safronov G.A. and Smironov G.N. (1993). LKr scintillation calorimeter. *Preprint ITEP 97*, Moscow.

Akimov D. Yu., Bolozdynya A.I., Churakov D.I., Lamkov V.A., Sadovsky A.A., Safronov G.A. and Smironov G.N. (1993). The influence of Xe doping on LKr scintillations. *Nucl Instr Meth A* 332: 575–576.

Akimov D., Belogurov S., Bolozdynya A., Churakov D., Chumakov M., Grishkin Yu., Pozdnyakov S. and Solovov V. (1994). Xenon scintillation drift chamber with microstrip readout. In: G. Della Mea and F. Sauli (eds). *Proc Intern Workshop on Micro-strip gas chambers, Legnaro (Italy), 13–14 October 1994*, Edizioni Progetto, Padova, 215–216.

Akimov D. Yu., Bolozdynya A.I., Churakov D.I., Lamkov V.A., Sadovsky A.A., Safronov G.A. and Smironov G.N. (1995). Scintillating LXe/LKr electromagnetic calorimeter. *IEEE Trans Nucl Sci* 42: 2244–2249.

Akimov D., Bewick A., Davidge D., Dawson J., Howard A.S., Ivaniouchenkov I., Jones W.G., Joshi J., Kudryavtsev V.A., Lawson T.B., Lebedenko V., Lehner M.J., Lightfoot P.K., Liubarsky I., Lüscher R., McMillan J.E., Peak C.D., Quenby J.J., Spooner N.J.C., Sumner T.J., *et al.* Measurements of scintillation efficiency and pulse shape for low energy recoils in liquid xenon. *Phys Lett B* 524: 245–251.

Aleksandrov Yu. A., Voronov G.S., Gorbunkov V.M., Delone N.B. and Nechaev Yu. I. (1967). *Bubble Chambers*, Indiana University Press, Bloomington and London.

Alkhazov D.G., Komar A.P. and Vorobiev A.A. (1967). *Nucl Instr Meth* 48: 1–12.

Alner G.J., Araújo H.M., Bewick A., Bungau C., Camanzi B., Carson M.J., Cashmore R.J., Chagani H., Chepel V., Cline D., Davidge D., Davies J.C., Daw E., Dawson J., Durkin T., Edwards B., Gamble T., Gao J., Ghag C., Howard A.S., Jones W.G., Josh M., Korolkova E.V., Kudryavtsev V.A., Lawson T., Lebedenko V.N., Lewin J.D., Lightfoot P.K., Lindote A., Liubarsky I., Lopes M.I., Lüscher R., Majewski P., Mavrokoridis K., McMillan J.E., Morgan B., Muna D., Murphy A. St. J., Neves F., Nicklin G.G., Ooi W., Paling S.M., Pinto da Cunha J., Plank S.J.S., Preece R.M., Quenby J.J., Robinson M., Salinas G., Sergiampietri F., Silva C., Solovov V.N., Smith N.J.T., Smith P.F., Spooner N.J.C., Sumner T.J., Thorne C., Tovey D.R., Tziaferi E., Walker R.J., Wang H., White J.T. and Wolfs F.L.H. (2007). First limits on WIMP nuclear recoil signals in ZEPLIN-II: A two phase xenon detector for dark matter detection. *Astropart Phys* 28: 287–302.

Amerio S., Amoruso S., Antonello M., Aprili P., Armenante M., Arneodo F., Badertscher A., Baiboussinov B., Baldo-Ceolin M., Battistoni G., Bekman B., Benetti P., Bernardini E., Bischofberger M., Borio di Tigliole A., Brunetti R., Bruzzese R.,

Bueno A., Calligarich E., Campanelli M., *et al.* (2004). Design, construction and tests of the ICARUS T600 detector. *Nucl Instr Meth A* 527: 329–410.

Ammosov V.V., Bolozdynya A.I., Kubantsev M.A., Lebedenko V.N. and Suvorov A.L. (1986). A new method of radiation detection in devices based on multi-channel plates. *Preprint ITEP*, ITEP: Moscow 48; *Pribory i Tehnika Experimenta* 6: 62–66 (in Russian).

Anciolotto F. and Toigo F. (1994). Properties of electron bubble approaching the surface of liquid helium. *Phys Rev B* 50: 12820–12830.

Anderson D.F., Charpak G., Holroyd R.A. and Lamb D.C. (1987). Liquid ionization chambers with electron extraction and multiplication in the gaseous phase. *Nucl Instrum Meth A* 261: 445–448.

Angle J., Aprile E., Arneodo F., Baudis L., Bernstein A., Bolozdynya A., Coelho L., Dahl E., Deviveiros L., Ferella A., Fernandes L., Fiorucci S., Gaitskell R J., Giboni K-L., Gomez R., Hasty R., Kwong J., Lopes J.A.M., Madden N., Manalaysay A., Manzur A., Mckinsey D., Monzani M.E., Ni K., Oberlack U., Orboeck J., Plante G., Santos J., Shagin P., Shutt T., Sorensen P., Winant C. and Yamashita M. (2007). 3D position sensitive XeTPC for dark matter search. *Nucl Phys B: Proc Supp* 173: 117–121.

Angle J., Aprile E., Arneodo F., Baudis L., Bernstein A., Bolozdynya A., Coelho L., Dahl E., Deviveiros L., Ferella A., Fernandes L., Fiorucci S., Gaitskell R.J., Giboni K-L., Gomez R., Hasty R., Kwong J., Lopes J.A.M., Madden N., Manalaysay A., Manzur A., Mckinsey D., Monzani M.E., Ni K., Oberlack U., Orboeck J., Plante G., Santos J., Shagin P., Shutt T., Sorensen P., Winant C. and Yamashita M. (2008a). First results from the XENON-10 dark matter experiment at the Gran Sasso National Laboratory. *Phys Rev Lett* 100: 1–5.

Angle J., Aprile E., Arneodo F., Baudis L., Bernstein A., Bolozdynya A., Coelho L.C.C., Dahl C.E., DeViveiros L., Ferella A.D., Fernandes L.M.P., Fiorucci S., Gaitskell R.J., Giboni K.L., Gomez R., Hasty R., Kastens L., Kwong J., Lopes J.A.M., Madden N., Manalaysay A., Manzur A., McKinsey D.N., Monzani M.E., Ni K., Oberlack U., Orboeck J., Plante G., Santorelli R., dos Santos J.M.F., Shagin P., Shutt T., Sorensen P., Schulte S., Winant C., and Yamashita M. XENON-10 Collaboration (2008b). Limits on Spin-Dependent WIMP–Nucleon Cross Sections from the XENON-10 Experiment, *Phys Rev Lett* 101.

Anisimov S.N., Barabash A.S., Bolozdynya A.I. and Stekhanov V.N. (1991). 'Mojdodyr' – apparatus for electric-spark purification of liquid krypton used in ionization detectors. *Instr and Exp Tech* 34(2): 313–316.

Anisimov S.N., Bolozdynya A.I. and Stekhanov V.N. (1984). Electron localisation and drift under the surface of condensed krypton. *JETP Letters* 40(3): 829–832 (in Russian).

Anisimov S.N., Bolozdynya A.I., Egorov V.V. and Stekhanov V.N. (1986). Application of mixtures of krypton and methane in emission detectors. *Preprint ITEP 86–104* Moscow: Atominform 8 (in Russian).

Emission Detectors

Anisimov S.N., Barabash A.S., Bolozdynya A.I. and Stekhanov V.N. (1989a). Control of electro-negative impurities contents in liquid krypton with emission detector. *Pribory i Tehnika Eksperimenta* 1: 79–82 (in Russian).

Anisimov S.N., Barabash A.S., Bolozdynya A.I. and Stekhanov V.N. (1989b). Measuring the ^{85}Kr content in krypton using a liquid ionization chamber. *Atomic Energy* 66: 415–417 (in Russian).

Aprile E. for XENON collaboration (2002). A 1-tonne liquid xenon experiment for a sensitive dark matter search. *Preprint astro-ph/0207670*.

Aprile E. and Baudis L. for XENON-100 collaboration (2008). Status and Sensitivity Projections for the XENON-100 Dark Matter Experiment. *e-print arXiv:0902*, 4253.

Aprile E., Giboni K.L., Majewski P., Ni K., Yamashita M., Gaitskell R., Sorensen P., DeViveiros L., Baudis L., Bernstein A., Hagmann C., Winant C., Shutt T., Kwong J., Oberlack U., McKinsey D. and Hasty R. (2005). The XENON dark matter search experiment. *New Astr Rev* 49: 289–295.

Aprile E., Bolotnikov A.E., Bolozdynya A.I. and Doke T. (2006). Noble Gas Detectors, Wiley-VCH Verlag GmbH & Co. KGaA, Weinheim.

Aprile E., Bolotnikov A., Chen D., Xu F. and Peskov V. (1994). First observation of the scintillation light from solid Xe, Kr and Ar with a CsI photo-cathode. *Nucl Instr Meth A* 353: 55–58.

Aprile E., Bolotnikov A., Chen D., Mukherjee R., Xu F., Anderson D.F., Peskov V. (1994). Performance of CsI photo-cathodes in liquid Xe, Kr, and Ar. *Nucl Instr Meth A* 338: 328–335.

Aprile E., Cushman P., Ni K. and Shagin P. (2005). Detection of liquid xenon scintillation light with a silicon photo-multiplier. *Nucl Instr Meth A* 556: 215–218.

Aprile E., Giboni K.L., Majewski P., Ni K., Yamashita M., Hasty R., Manzur A. and McKinsey D.N. (2005). Scintillation response of liquid xenon to low energy nuclear recoils. *Phys Rev D* 72.

Aprile E., Giboni K.L., Majewski P., Ni K. and Yamashita M. (2004). Proportional light in a dual-phase xenon chamber. *IEEE Trans Nucl Sci* 51: 1986–1990.

Aprile E., Giboni K.L. and Rubbia C. (1985). A study of ionization electrons drifting large distances in liquid and solid argon. *Nucl Instr Meth A* 241: 62–71.

Aprile E., Oberlack U.G., Curioni A., Egorov V., Giboni K.L., Ventura S., Doke T., Kikuchi J., Takizawa K., Chupp E .L. and Dunphy P.P. (2000). Spectroscopy and imaging performance of the Liquid xenon Gamma-Ray Imaging Telescope (LXeGRIT). *Proc SPIE* 4140: 333–343.

Araújo H.M., Chepel V., Lopes M.I., van der Marel J., Ferreira Marques R. and Policarpo A.J.P.L. (1998). Study of bialkali photo-cathodes below room temperature in the UV/VUV region. *IEEE Trans Nucl Sci* 45: 542–549.

Araújo H.M., Bewick A., Davidge D., Dawson J., Ferbel T., Howard A.S., Jones W.G., Joshi M., Lebedenko V., Liubarsky I., Quenby J.J., Sumner T.J. and Neves F. (2004). Low-temperature study of 35 photo-multiplier tubes for the ZEPLIN III experiment. *Nucl Instr Meth A* 521: 407–415.

Arisaka K., Wang H., Smith P.F., Cline D., Teymourian A., Brown E., Ooi W., Aharoni D., Lam C.W., Lung K., Davies S. and Price M. (2009). XAX: A multi-ton, multi-target detection system for dark matter, double-beta decay and *pp* solar neutrinos. *Astropart Phys* 31: 63–74.

Arneodo F., Baiboussinov B., Badertscher A., Benetti P., Bernardini E., Bettini A., Borio di Tiogliole A., Brunetti R., Bueno A., Calligarich E., Campanelli M., Carpanese C., Cavalli D., Cavanna F., Cennini P., Centro S., Cesana A., Cline D., De Mitri I., Dolfini R. *et al.* Scintillation efficiency of nuclear recoil in liquid xenon. *Nucl Instr Meth A* 449: 147–157.

Arodzero A., Bolozdynya A., Bolotnikov A., Proctor A. and Richards J. (2004). Two-channel high-pressure helium-3 scintillation neutron detector. *IEEE Trans Nucl Sci* 51: 322–327.

Arvanov A.N., Akhperdzhanyan A.G. and Gavalyan V.G. (1981). Detector with controlled secondary electron emission working with MgO. *Pribory i Tehnika Eksperimenta* 4: 58–61 (in Russian).

Ascarelli G. (1979). The role of shallow traps on the mobility of electrons in liquid Ar, Kr, and Xe. *J Chem Phys* 71: 5030–5033.

Atkins K.R. (1959). Ions in liquid helium. *Phys Rev* 116: 1339–1343.

Atrazhev V.M., Berezhnov A.V., Dunikov D.O., Chernysheva I.V., Dmitrenko V.V. and Kapralova G. (2005). Electron transport coefficients in liquid xenon. *Proc 2005 IEEE Int Conf Diel Liquids Port* June 26–July 1: 329–332.

Badoz J., Le Liboux M., Nahoum R., Israel G., Raulin F. and Torre J.P. (1992). A sensitive cryogenic refractometer: Application to the refractive index determination of pure or mixed liquid methane, ethane, and nitrogen *Rev Sci Inst* 63: 2967–2973.

Bakale G., Sowada U. and Schmidt W.F. (1976). Effect of an electric field on electron attachment to SF_6, N_2O, and O_2 in liquid argon and xenon. *J Phys Chem* 80: 2556–2559.

Balakin A.A., Boriev I.A. and Yakovlev B.S. (1977). Thermal emission of excess electrons from liquid hydrocarbons. *Can J Chem* 55: 1985–1986.

Balakin A.A., Boriev I.A. and Yakovlev B.S. (1978). Thermal emission of excess electrons from liquid hydrocarbons. *High Energy Chem* 12: 20–22 (in Russian).

Baldini G. (1962). Ultraviolet Absorption of Solid Argon, Krypton, and Xenon. *Phys Rev* 128: 1562–1567.

Baldini A., Bemporad C., Cei F., Doke T., Grassi M., Grebenyuk A., Grigoriev D., Haruyama T., *et al.* (2005). Absorption of scintillation light in a 100 l liquid xenon γ-ray detector and expected detector performance. *Nucl Instr Meth A* 545: 753–764.

Baldo-Ceolin M., Puglierin G., Barmin V., Barabash A., Bolozdynya A., Dolgolenko A., Nozik V., Starostin A., Shebanov V., Tumanov G. and Zeldovich O. (1992). Large volume liquid Xe detector for $\overline{V}_e$ magnetic moment measurement from $\overline{V}_e e$ scattering at nuclear power reactor. *Preprint ITEP Mosc* 35–92.

Balzer F., Kankate L., Niehus H. and Rubahn H.-G. (2005). Nanoaggregates from oligothiophenes and oligophenylenes – a systematic growth survey. *Proc SPIE* 5724: 285–294.

Bandis C. and Pate B.B. (1996). Simultaneous field emission and photo-emission from diamond. *Appl Phys Lett* 69(3): 366–368.

Barabanov I.R., Gavrin V.N., Girin S.V., Kornoukhov V.N. and Pshukov A.M. (1986). Search for double-beta decay of ^{136}Xe. *JETP Letters* 43: 166–167 (in Russian).

Barabanov I.R., Belli P., Bernabei R., Dai C.J., Ding L.K., Di Nicolantonio W., Guretsov V.I., Ianovich E.A., Incicchitti A., Janz V.E., Kornoukhov V.N., Kuang H.H., Ma J.M., Montecchia F., Orekhov I.V., Danshin C.V. and Prosperi D. (1997). Perspectives to measure neutrino magnetic moment deep underground. *Astropart Phys* 8: 67–76.

Barabash A.S. and Bolozdynya A.I. (1993). *Liquid Ionization Detectors*, EnergoAtomizdat, Moscow (in Russian).

Barabash A.S. and Bolozdynya A.I. (1989). How to detect the dark matter of the galaxy if it is made up of weakly interacting neutral particles with masses 1–10 GeV/c^2. *JETP Lett* 49: 356–359.

Barabash A.S., Golubev A.A., Kazachenko O.V. and Ovchinnikov B.M. (1985). Investigation of electronic conductivity of liquid argon–nitrogen mixtures. *Nucl Instr Meth A* 234: 451–454.

Barkov L.M., Grebenuk A.A., Ryskulov N.M., Stepanov Yu. P. and Zverev S.G. (1996). Measurement of the refractive index of liquid xenon for intrinsic scintillation light, *Nucl Instr Meth A* 379: 482–483.

Barmin V.V., Barylov V.G., Golubchikov V.M., Gorokhov A.I., Demidov V.S., Konoplev N.S. and Shebanov V.A. (1972). Electron energy measurement in xenon bubble chamber by shower tracks total length method. *Preprint ITEP-938* (in Russian).

Barmin V.V., Borisov V.S., Davidenko G.V., Dolgolenko A.G., Guaraldo C., Larin I.F., Matveev V.A., Petrascu C., Shebanov V.A., Shishov N.N., Sokolov L.I. and Tumanov G.K. (2003). Observation of a baryon resonance with positive strangeness in K^+ collisions with Xe nuclei. *Phys At Nucl* 66: 1715–1718.

Belogurov S., Bolozdynya A., Churakov D., Koutchenkov A., Morgunov V., Solovov V., Safronov G. and Smirnov G. (1995). High pressure gas scintillation drift chamber with photo-multipliers inside working medium. *1995 IEEE NSS and MIC Record vol. 1 San Fran* October 21–28 1995: 519–523.

Beda A.G., Demidova E.V., Starostin A.S., Brudanin V.B., Egorov V.G., Medvedev D.V., Shirchenko M.V., Vylov Ts. (1999). GEMMA experiment: three years of the search for the neutrino magnetic moment. *e-print arXiv:0906.1926v1 [hep-ex]* 10 June 2009.

Benetti P., Montanari C., Raselli G.L., Rossella M. and Vignoli C. (2003). Detection of the VUV liquid argon scintillation light by means of glass-window photo-multiplier tubes. *Nucl Instrum Meth A* 505: 89–92.

Benetti P., Acciarri R., Adamo F., Baibussinov B., Baldo-Ceolin M., Belluco M., Calaprice F., Calligarich E., Cambiaghi M., Carbonara F., Cavanna F., Centro S., Cocco A.G., Di Pompeo F., Ferrari N., Fiorillo G., Galbiati C., Gallo V., Grandi L., Ianni A., Mangano G., Meng G., Montanari C., Palamara O., Pandola L., Pietropaolo F., Raselli G.L., Rossella M., Rubbia C., Szelc A.M., Ventura S. and

Vignoli C. WARP Collaboration (2008). First results from a dark matter search with liquid argon at 87 K in the Gran Sasso Underground Laboratory, *Astropart Phys* 28: 495–507.

Bernabei R., Belli P., Cappella F., Cerulli R., Dai C.J., d'Angelo A., He H.L., Incicchitti A., Kuang H.H., Ma J.M., Montecchia F., Nozzoli F., Prosperi D., Sheng X.D. and Ye Z.P. (2008). First results from DAMA/LIBRA and the combined results with DAMA/NaI. *Eur Phys J C* 56: 333–355.

Berset J.C., Burns M., Geissler K., Harigel G., Lindsay J., Linser G., and Schenk F. (1982). Scintillation light from liquid argon and its use in a new hybrid detector. *Nucl Instr Meth* 203: 133–140.

Blinov G., Krestnikov Yu. and Pershin I. (1954). Observation of tracks of high-energy particles from accelerator with propane bubble chamber. *Dokl Acad Sci USSR* XCIX: 929–930 (in Russian).

Bolozdynya A. (1999). Two-phase emission detectors and their applications. *Nucl Instrum Meth A* 422: 314–320.

Bolozdynya A. (2006). Two-phase electron emission detectors. *IEEE Trans Diel Electr Insul* 13: 616–623.

Bolozdynya A.I. (1985). Emission detectors of particles. *Приборы и Техника Эксперимента* 2: 5–23; *Instr Exp Tech* 28(2): 265–396.

Bolozdynya A.I. (1986). To electron emission from liquid iso-octane. *Preprint ITEP*, 86–103 (in Russian).

Bolozdynya A.I. (1986). Excess electron emission from condensed krypton and other non-polar dielectrics. *Preprint ITEP Mosc Atominform.* 172–86 (in Russian).

Bolozdynya A.I (1991). Transport of excess electrons through and along condensed krypton interface. *Proc 3rd Int Conf Porp and Appl Diel Mat July 8–12 1991, Tokyo* 841–844.

Bolozdynya A.I., Bradley A.W., Brusov P.P., Dahl C.E., Kwong J. and Shutt T. (2008). Using a wavelength shifter to enhance the sensitivity of liquid xenon dark matter detectors. *IEEE Trans Nucl Sci* 55: 1453–1457.

Bolozdynya A.I., Brusov P.P., Shutt T., Dahl C.E. and Kwong J. (2007). A chromatographic system for removal of radioactive 85Kr from xenon. *Nucl Instr Meth A* 579: 50–53.

Bolozdynya A.I., Egorov O.K., Korshunov A.A., Sokolov L.I., Miroshnichenko V.P., and Rodionov B.U. (1977). The first observations of particle tracks in condensed matter obtained by the emission method. *Pis'ma Zhurnal Eksp Teor Fiz*, 25: 401–404 (in Russian).

Bolozdynya A.I., Egorov O.K., Miroshnichenko V.P., Rodionov B.U. and Shuvalova E.N. (1980). A new possibility to search for low-ionizing particles. In: *Elementarnye Chastitsy i Kosmicheskie Luchi 5*, pp. 65–72 Atomizdat, Moscow (in Russian).

Bolozdynya A.I., Egorov O.K., Sokolov L.I., Miroshnichenko V.P. and Rodionov B.U. (1980). Solid Krypton Emission Chamber. In: *Solid State Nuclear Track Detectors*, H. Francois *et al.* (eds), pp. 29–32, Oxford–New York, Pergamon Press.

Bolozdynya A., Egorov V., Koutchenokov A., Safronov G., Smirnov G., Medved S. and Morgunov V. (1997a). A high pressure xenon self-triggered scintillation drift chamber with 3D sensitivity in the range of 20–140 keV deposited energy. *Nucl Instrum Meth A* 385: 225–238.

Bolozdynya A., Egorov V., Koutchenokov A., Safronov G., Smirnov G., Medved S. and Morgunov V. (1997b). An electroluminescence emission detector to search for double-beta positron decays of ^{134}Xe and ^{78}Kr. *IEEE Trans Nucl Sci* 44: 1046–1051.

Bolozdynya A.I., Egorov V.V., Kalashnikov S.D., Krivoshein L., Miroshnichenko V.P. and Rodionov B.U. (1981). Detector of coordinates for low energy particles based on condensed krypton. *Preprint ITEP-113*, Moscow: Institute for Theoretical and Experimental Physics (in Russian).

Bolozdynya A.I., Egorov V.V., Kalashnikov S.D., Krivoshein L., Miroshnichenko V.P. and Rodionov B.U. (1982). Emission electroluminescence gamma camera based on condensed xenon, *Preprint ITEP-37*, Moscow: Institute for Theoretical and Experimental Physics (in Russian).

Bolozdynya A.I., Egorov V.V., Kalashnikov S.D., Krivoshein L., Miroshnichenko V.P. and Rodionov B.U. (1985). Emission electroluminescence chamber with condensed xenon working medium. *Pribory i Tehnika Eksperimenta* 4: 43–45 (in Russian).

Bolozdynya, A., Egorov, V., Rodionov, B., Miroshnichenko, V. (1995). Emission detectors, *IEEE Trans. Nucl. Sci.* 42: 565-569.

Bolozdynya A.I., Lebedenko V.N., Rodionov B.U., Balakin A.A., Boriev I.A., Yakovlev B.S. (1978). Electrostatic emission of electrons into the gas phase from alpha-tracks in the liquid iso-octane. *Zhurnal Tehnicheskoy Fiziki* 48: 1514–1519 (in Russian).

Bolozdynya A.I., Miroshnichenko V.P. and Rodionov B.U. (1977). Electrostatic emission of free electrons from liquid and solid argon. *Pis'ma Zhurnal Tehnicheskoj Fiziki* 2: 64–67 (in Russian).

Bolozdynya A.I. and Stekhanov V.N. (1984). Capture of quasi-free electrons in liquid krypton. *Preprint ITEP Mosc* 27 (in Russian).

Boisvwert J C., Montroy J.T., Jostad L., Zhou B. and Szawlowski M. (1996). Improved large-area avalanche photo-diodes for scintillation detection in calorimetry. In: *1996 IEEE Nuclear Science Symposium Conference Record, November 2–9, 1996, Anaheim, California* 1: 16–20.

Bond L., Collar J.I., Ely J., Flake M., Hall J., Jordan D., Nakazawa D., Raskin A., Sonnenschein A. and O'Sullivan K. (2005). Development of bubble chambers with sensitivity to WIMPs. *Nucl Phys B (Proc Suppl)* 138: 68–71.

Bondar A., Buzulutskov A., Shekhtman L., Snopkov R. and Tikhonov Yu. (2004). Cryogenic avalanche detectors based on gas electron multipliers. *Nucl Instrum Meth A* 524: 130–141.

Bondar A., Buzulutskov A., Grebenuk A., Pavlyuchenko D., Snopkov R. and Tikhonov Yu. (2006). Two-phase argon and xenon avalanche detectors based on Gas Electron Multipliers. *Nucl Instrum Meth A* 556: 273–280.

Bondar A., Buzulutskov A., Grebenuk A., Pavlyuchenko D., Snopkov R. and Tikhonov Yu. (2007). First results of the two-phase argon avalanche detector performance with CsI photo-cathode. *e-print www.arxiv.org/0702237*.

Boors H.A. and Motz L.L. (1966). *The world of the atom*, Basic Books, New York.

Boriev I.A., Balakin A.A. and Yakovlev B.S. (1978). Electron emission from non-polar liquids. *Khimiya Vysokih Energiy* 12: 20–25 (in Russian).

Borghesani A.F. (2007). Ions and electrons in liquid helium. In: *International series of monographs on physics 137* Oxford University Press, Oxford.

Borghesani A.F., Carugno G., Cavenago M. and Conti E. (1990). Electron transmission through the Ar liquid–vapor interface. *Phys Lett A* 149(9): 481–484.

Borghesani A.F., Carugno G. and Santini M. (1991). Experimental determination of the conduction band of excess electrons in liquid argon. *IEEE Trans Elect Insul* 26: 615–622.

Borghesani A.F., Iannuzzi D. and Carugno G. (1997). Excess electron mobility in liquid Ar–Kr and Ar–Kr, *J Phys: Condens Matter* 9: 5057–5065.

Boutboul T., Akkerman A., Giberkhterman A., Breskin A. and Chechik R. (1999). An improved model for ultra-violet- and X-ray-induced electron emission from CsI. *J Appl Phys* 86: 5841–5849.

Boyle F.P. and Dahm A.J. (1976). Extraction of charged droplets from charged surfaces of liquid dielectrics. *J L Temp Phys* 23: 477–486.

Braem A., Gonidec A., Schinzel D., Sedel W., Clayton E.F., Davies G., Hall G., Payne R., Roe S., Seez C., Striebig J., Virdee T.S. and Cockerill D.J.A. (1992). Observation of the UV scintillation light from high energy electron showers in liquid xenon. *Nucl Instrum Meth A* 320: 228–237.

Bronic I.K. (1992). On a relation between the W value and the Fano factor. *J Phys B* 25: L215–L218.

Brown J.L., Glaser D. and Perl M. (1956). Liquid xenon bubble chamber. *Phys Rev* 102: 586–587.

Brown J.L., Bryant H.C., Burnstein R.A., Glaser D., Hartung R., Kadyk J.A., VanPuttent J.D., Sinclair D., Trilling G.H. and Van der Velde J.C. (1961). Properties of neutral strange particles produced in a xenon bubble chamber. *Nuovo Cimento* XIX: 1155–1170.

Brown T.R. and Grimes C.C. (1972). Observation of Cyclotron Resonance in Surface-bound Electrons on Liquid Helium. *Phys Rev Lett* 29: 1233–1236.

Bruschi L., Maravigkia B. and Moss F.E. (1966). Measurement of a Barrier for the Extraction of Excess Electrons from Liquid Helium. *Phys Rev Lett* 17: 682–684.

Bruschi L., Mazzi G. and Torzo G. (1975). Transmission of negative ions through the liquid vapour surface in neon. *J Phys C: Solid State Phys* 8: 1412–1422.

Bugg D. (1959). Bubble chamber. *Progr Nucl Phys* 7: 1–25.

Buzhan P., Dolgoshein B., Ilyin A., Kantzerov V., Kaplin V., Karakash A., Plesco A., Popova E., Smirnov S. and Volkov Yu. (2001). An advanced study of silicon photo-multiplier. *ICFA Instrum Bull* 23: 28–34.

Buzhan P., Dolgoshein B., Filatov L., Ilyin A., Kantzerov V., Kaplin V., Karakash A., Kayumov F., Klemin S., Popova E. and Smirnov S. (2003). Silicon photomultiplier and its possible applications. *Nucl Instrum Meth A* 504: 48–52.

Buzulutskov A., Breskin A., and Chechik R. (1995). Field enhancement of the photoelectric and secondary electron emission from CsI. *J Appl Phys* 77: 2138–2145.

Cabrera B., Clarke R.M., Colling P., Miller A.J., Nam S. and Romani R.W. (1998). Detection of single infrared, optical, and ultra-violet photons using superconducting transition edge sensors. *Appl Phys Lett* 73: 737–737.

Careri G., Fasoli U. and Gaeta F.S. (1960). Experimental behaviour of ionic structures in liquid helium-II. *Nuovo Cimento* 15: 774–783.

Carugno G. (1998). Infrared emission in gaseous media induced by ionizing particles and by drifting electrons. *Nucl Instrum Meth A* 419: 617–620.

Chen M., Mullins M., Pelly D., Shotkin S., Sumorok K., Akyuz D., Chen E., Gaudreau M.P.J., Bolozdynya A., Tchernyshev V., Goritchev P., Khovansky V., Koutchenkov A., Kovalenko A., Lebedenko V., Vinogradov V., Gusev L., Sheinkman V., Krasnokutsky R.N., Shuvalov R.S. *et al.* (1993). Homogeneous scintillating LKr/Xe calorimeters. *Nucl Instrum Meth A* 327: 187–192.

Chung M.S. and Yoon B.-G. (2003). Analysis of the slope of the Fowler–Nordheim plot for field emission from n-type semiconductors. *J Vac Sci Tech B* 21(1): 548–51.

Cline D. and Hong W. (1992). A liquid argon or xenon detector to observe a neutrino magnetic moment of $\mu_\nu \sim 10^{-11}\,\mu_B$. *Int J Mod Phys A* 7: 4167–4173.

Collar J.I., Puibasset J., Girard T.A., Limagne D., Miley H.S. and Waysand G. (2000). First dark matter limits from a large-mass, low-background superheated droplet detector. *Phys Rev Lett* 85: 3083–3086.

Cohen M.H. and Lekner J. (1967). Theory of hot electrons in gases, liquids and solids. *J Phys Rev* 158: 305–309.

Conde C.S.N. (2004). Gas proportional scintillation counters for X-ray spectrometry. In: *X-Ray Spectrometry: recent technological advances*, Tsuji K., Injuk J. and Van Greiken R. (eds), pp. 195–216, John Wiley & Sons, USA.

Croxton C.A. (1974). Liquid State Physics: A Statistical Mechanical Introduction. In: *Series: Cambridge Monographs on Physics*, Cambridge University Press, Cambridge.

Davidson N. and Larsh A.E. (1948). Conductivity pulses in liquid argon. *Phys Rev* 74: 220–220.

Davies G.J., Spooner N.J.C., Davies J.D., Pyle G.J., Bucknell T.D., Squier G.T.A., Lewin J.D. and Smith P.F. (1994). The scintillation efficiency for calcium and fluorine recoils in CaF_2 and carbon and fluorine recoils in C_6F_6 for dark matter searches. *Phys Lett B* 322: 159–165.

Derenzo S.E., Mast T.S., Zaklad H. and Muller R.A. (1974). Electron avalanche in liquid xenon. *Phys Rev A* 9: 2582–2591.

DeSalvo R., Hao W., You K., Wang Y. and Xu C. (1992). First results on the hybrid photo-diode tubes. *Nucl Instrum Meth A* 315: 375–384.

Dias T.H.V.T., Rachinhas P.J.B.M., Lopes J.A.M., Satos F.P., Tavora L.M.N., Conde C.A.N. and Stauffer A.D. (2004). The transmission of photo-electrons emitted from CsI photo-cathodes into Xe, Ar, Ne and their mixtures: A Monte Carlo study of the dependence on E/N and incident VUV photon energy. *J Phys D Appl Phys* 37: 540–549.

Doke T. (1981). Fundamental properties of liquid argon, krypton and xenon as radiation media. *Portugal Phys* 12: 9–48.

Doke T. (2005). Ionization and excitation by high-energy radiation. In: *Electronic excitations in liquefied rare gases*, Schmidt W.F. and Illenberger E. (eds), pp. 71–93, American Scientific Publishers, California.

Doke T. and Masuda K. (1999). Present status of liquid rare gas scintillation detectors and their new application to gamma-ray calorimeters. *Nucl Instrum Meth A* 420: 62–80.

Dolgoshein B.A., Lebedenko V.N. and Rodionov B.U. (1967). Electric field luminescence of liquid xenon activated with alpha-particles. *ZhETF Pis'ma*, 6: 755–757 (in Russian).

Dolgoshein B.A., Lebedenko V.N. and Rodionov B.U. (1970). New method of registration of ionizing particle tracks in condensed matter. *JETP Lett* 11: 351–353.

Dolgoshein B.A., Lebedenko V.N. and Rodionov B.U. (1973). Some electron methods of detection of particle tracks in liquids. *Elem Part and Cosm Rays* 2: 86–91 (in Russian).

Dolgoshein B.A., Kruglov A.A., Lebedenko V.N., Miroshnichenko V.P. and Rodionov B.U. (1973). Electronic method of registration of particles in two-phase liquid-gas systems. *Fizika Elemementanyh Chastits i Atomnogo Yadra* 4: 167–186 (in Russian).

Egorov O.K. and Stepanov A.M. (2000). A universal Marx generator with double shielding. *Instr and Exp Tech* 43: 339–344; *Pribory I Tekhinika Eksperimenta* 3: 61–66 (in Russian).

Egorov V.V., Miroshnichenko V.P., Rodionov B.U., Bolozdynya A.I., Kalashnikov S.D. and Krivoshein V.L. (1983). Electroluminescence emission gamma-camera. *Nucl Instrum Meth* 205: 373–374.

Elster J. and Geitel H. (1890). The use of sodium amalgam in photo-electric experiments. *Ann Physik* 41: 161–165.

Ellis J., Olive K.A., Santoso Y. and Spanos V.C. (2005). Update on the direct detection of supersymmetric dark matter. *Phys Rev D* 71: 095007.

Engel J. (1991). Nuclear form factors for the scattering of weakly interacting massive particles. *Phys Lett B* 264: 114–119.

Ereditato A. and Rubbia A. (2005). Ideas for suture liquid argon detectors. *Nucl Phys B Proc Suppl* 139: 301–310.

Ereditato A. and Rubbia A. (2006). The liquid argon TPC: A powerful detector for future neutrino experiments and proton decay searches. *Nucl Phys B Proc Suppl* 154: 163–178.

Fabian C.W. and Gianotti F. (2003). Calorimetry for particle physics. *Rev Mod Phys* 75: 1243–1284.

Gavalyan V.G., Lorikyan M.P., and Arvanov A.N. (1983). Multiplication of electrons in emitters of controllable secondary electron emission (CSEE). *Zhurnal Tehnicheskoj Fiziki* 53: 1621–1624 (in Russian)

Gerbier G., Lesquoy E., Rich J., Spiro M., Tao C., Yvon D., Zylberajch S., Delbourgo P., Haouat G., Humeau C., Goulding F., Landis D., Madden N., Smith A., Walton J., Caldwell D.O., Magnusson B., Witherell M., Sadoulet B. and Da Silva A. (1990). Measurement of the ionization of slow silicon nuclei in silicon for the calibration of a silicon dark-matter detector. *Phys Rev D* 42: 3211– 3214.

Glaser D.A. (1952). Some effects of ionizing radiation on the formation of bubbles in liquids. *Phys Rev* 87: 665.

Gorodkov Yu. B., Lyubimov V.A., Sidorov I.V. and Soloschenko V.A. (1974). Cryogenic tracking spark chamber. *Pribory I Tekhinika Eksperimenta* 6: 46–47 (in Russian).

Grand D. and Bernas A. (1977). The ionization potential of a solute and the ground state energy of the excess electron. *J Phys Chem* 81: 1209–1211.

Grandy L. (2005). WARP: An argon double phase technique for dark matter search. PhD thesis, University of Pavia, Pavia.

Gryko J and Popielawski J. (1977). Comment on the application of the Cohen-Lekner theory to excess electron mobility in liquid krypton. *Phys Rev* 16: 1333–1336.

Gurvich L.V., Karachentsev G.B. and Kondratiev V.N. (1974). Breaking energy of chemical bonds. In: *Ionization potential and electron affinity*, Nauka, Moscow (in Russian).

Guschin E.M. (1981). Investigation of electron dynamics in position-sensitive detectors based on condensed argon and xenon. PhD thesis, MEPI, Moscow (in Russian).

Guschin E.M., Kruglov A.A., Litskevich V.V., Lebedev A.N., Obodovsky I.M. and Somov S.V. (1979). Electron emission from condensed noble gases. *Zhurnal Eksperimental'noj i Teoreticheskoy Fiziki* 76: 1685–1689 (in Russian).

Guschin E.M., Kruglov A.A., Obodovsky I.M. (1982). Dynamics of electrons in condensed argon and xenon. *Zhurnal Eksperimental'noj i Teoreticheskoy Fiziki* 82: 1114–1125 (in Russian).

Guschin E.M., Kruglov A.A. and Obodovsky I.M. (1982). Process of 'hot' electron emission from liquid and solid argon and xenon. *Zhurnal Eksperimental'noj i Teoreticheskoy Fiziki* 82: 1485–1490 (in Russian).

Guschin E.M., Kruglov A.A., Obodovsky I.M., Pokachalov S.G. and Shilov V.A. (1982). Liquid xenon position-sensitive detector of gamma radiation. *Pribory i Tehnika Eksperimenta* 3: 49–52 (in Russian).

Hagmann C. and Bernstein A. (2004). Two-phase emission detector for measuring coherent neutrino-nucleus scattering. *IEEE Trans Nucl Sci* 51: 2151–2155.

Harigel G.G. (2003). Bubble chambers, technology and impact on high energy physics. In: *Bologna 2003, 30 years of bubble chamber physics*, pp. 342–355, INFN, Bologna, Italy.

Hayashi T. (1982). Recent developments in photo-multipliers for nuclear radiation detectors. *Nucl Instrum Meth* 196: 181–186.

Hayden M.E., Archibald G., Barnes P.D., Buttler W.T., Clark D.J., Cooper M.D., Espy M., Golub R., Greene G.L., Lamoreaux S.K., Lei C., Marek L.J., Peng J.-C. and Penttila S.I. (2004). Neutron-detected Tomography of Impurity-Seeded Superfluid Helium. *Phys Rev Lett* 93.

Henderson C. (1970). *Cloud and bubble chambers*, Butler & Tanner Ltd, London.

Herb G K. and Van Sciver W.J. (1965). Measurement of the Decay Time of Sodium Salicylate. *Rev Sci Instrum* 36: 1650–1652.

Hertz H. (1887). Ultra-violet light and electric discharge. *Ann Physik* 31: 983–1000.

Hilt O., Schmidt W.F., Khrapak A.G. (1994). Ionic mobilities in liquid xenon. *IEEE Trans Diel Elect Insul* 1: 648–656.

Hitachi A. (2007). Quenching factor and electronic LET in a gas at low energy. *J Phys Conf Ser* 65.

Hitachi A. and Takahashi T. (1983). Effect of ionization density on the time dependence of luminescence from liquid argon and xenon. *Phys Rev B* 27: 5279–5285.

Hitachi A., Yunoki A., Doke T. and Takahashi T. (1987). Scintillation and ionization yield for α particles and fission fragments in liquid argon. *Phys Rev A* 35: 3956–3958.

Hori M. (2004). Parallel plate chambers for monitoring the profiles of high-intensity pulsed antiproton beams. *Nucl Instr Meth A* 522: 420–431.

Hu W.S., Lin Y.F., Tao Y.T., Hsu Y.J. and Wei D.H. (2005). Highly oriented growth of p-sexiphenyl molecular nanocrystals on rubbed polymethylene surface. *Macro Molecules* 38: 9617–9624.

Huang S.-S. and Freeman G.R. (1977). Electron mobilities in gaseous, critical, and liquid xenon: Density, electric field, and temperature effects: Quasi-localisation. *J Chem Phys* 68(4): 1355–1362.

Huang Y.H., Lanou R.E., Maris H.J., Seidel G.M., Sethumadhavan B. and Yao W. (2008). Potential for precision measurement of solar neutrino luminosity by HERON. *Astropart Phys* 30: 1–11.

Huxley L.G.H. and Crompton R.W. (1974). *The diffusion and drift of electrons in gases*, John Wiley & Sons, New York–London–Sydney–Toronto.

Hutchinson G.W. (1948). Ionization in liquid and solid argon. *Nature* 162: 610–611.

Ichige M., Doke T., Doi Y. and, Yoshimura Y. (1993). Operating characteristics of photo-multipliers at low temperature. *Nucl Instrum Meth A* 327: 144–147.

Ishida N., Doke T., Kikuchi J., Kuwahara K., Kashiwagi T., Ichige M., Hasuike K., Ito K., Ben S., Hitachi A., Qu Y.H., Masuda K., Suzuki M., Kase M., Takahashi T., Chen M., Sumorok S., Gaudreau M. and Aprile E. (1993). Measurement of attenuation length of scintillation light in liquid xenon. *Nucl Instrum Meth A* 327: 152–154.

Kadkhoda P., Ristau D. and von Alvensleben F. (1998). Total Scatter Measurements in the DUD/VUV. *Proc SPIE* 3578: 544–554.

Kalibjian R. (1965). A photo-tube using a semiconductor diode as a multiplier element. *IEEE Trans Nucl Sci* 12: 367–369.

Kaufmann L. and Rubbia A. (2007). The ArDM project: A Direct Detection Experiment, Based on Liquid Argon, for the Search of Dark Matter, *Nucl Phys B – Proc Suppl* 173: 141–143.

Khaikin M.S. and Volodin A.P. (1978). Disturbance of stability of the charged liquid helium surface and formation of bubblons. *Sov Phys Usp* 21(12): 1006–1007.

Kalashnikov S.D. (1985). *Basic physics of scintillation gamma camera design*, EnergoAtomizdat, Moscow (in Russian).

Khrapak A.G., Schmidt W.F. and Illenberger E. (2005). Localised electrons, holes and ions. In: *Electronic excitations in liquefied rare gases*, Schmidt W.F. and Illenberger E. (eds), pp. 239–273, American Scientific Publishers, California.

Kim J.G., Dardin S.M., Kadel R.W., Kadyk J.A., Peskov V. and Wenzel W.A. (2004). Electron avalanches in liquid argon mixtures. *Nucl Instrum Meth A* 534: 376–396.

Kirilenov A.V. and Konovalov S.P. (1981). Xenon of high density in detectors for ionizing radiation. *Preprint of Lebedev's Physical Institute Mosc* 149 (in Russian).

Kobayashi S., Mizusawa T., Shirai K., Saito Y., Ogiwara N. and Latham R.V. (1999). Electron emission microscope for use in electron emission sites observations. *Appl Surface Sci* 146: 148–151.

Lally C.H. (1994). UV quantum efficiency of organic fluors. *Imperial College Internal Note IC/HEP/94-11*.

Landau L.D. and Lifshitz E.M. (1984). *Electrodynamics of Continuous Media*, 2nd Edition, Pergamon, Oxford.

Lansiart A., Seigneur A., Moretti J.-L. and Morucci J.P. (1976). Development research on a highly luminous condensed xenon scintillator. *Nucl Instrum Meth* 135: 47–52.

Lebedenko V.N., Araujo H.M., Barnes E.J., Bewick A., Cashmore R., Chepel V., Davidge D., Dawson J., Durkin T., Edwards B., Ghag C., Graffagnino V., Horn M., Howard A.S., Hughes A.J., Jones W.G., Joshi M., Kalmus G.E., Kovalenko A.G., Lindote A., Liubarsky I., Lopes M.I., Luscher R., Lyons K., Majewski P., Murphy A. St. J., Neves F., Pinto da Cunha J., Preece R., Quenby J.J., Scovell P.R., Silva C., Solovov V.N., Smith N.J.T., Smith P.F., Stekhanov V.N., Sumner T.J., Thorne C. and Walker R.J. (2008). Result from the First Science Run of the ZEPLIN-III Dark Matter Search Experiment. *e-print arXiv:0812.1150*.

Lebedenko V.N., Araujo H.M., Barnes E.J., Bewick A., Cashmore R., Chepel V., Davidge D., Dawson J., Durkin T., Edwards B., Ghag C., Graffagnino V., Horn M., Howard A.S., Hughes A.J., Jones W.G., Joshi M., Kalmus G.E., Kovalenko A.G., Lindote A., Liubarsky I., Lopes M.I., Luscher R., Lyons K., Majewski P., Murphy A. St. J., Neves F., Pinto da Cunha J., Preece R., Quenby J.J., Scovell P.R., Silva C., Solovov V.N., Smith N.J.T., Smith P.F., Stekhanov V.N., Sumner T.J., Thorne C. and Walker R.J. (2009). Limits on the spin-dependent WIMP-nucleon cross-sections from the first science run of the ZEPLIN-III experiment. *e-print arXiv:0901.4348v1*.

Le Comber P.G., Loveland R.J. and Spear W.E. (1975). Hole transport in the rare-gas solids Ne, Ar, Kr, and Xe. *Phys Rev B* 11: 3124–3130.

Lekner J. (1967). Motion of electrons in liquid argon. *Phys Rev* 158: 130–137.

Lekner J. (1968). Mobility maxima in the rare-gas liquids. *Phys Lett. A* 27: 341–342.

Levine J.L. and Sanders T.M. (1967). Mobility of Electrons in Low-Temperature Helium Gas. *Phys Rev* 154: 138–149.

Lewin J.D. and Smith P.F. (1996). Review of mathematics, numerical factors, and corrections for dark matter experiments based on elastic nuclear recoil. *Astropart Phys* 6: 87–112.

Lindhard J., Nielsen V., Scharf M. and Thomsen P.M. (1963). Integral equations covering radiation effects. *Mater Fys Med Dan Vid Selsk* 33: 1–9.

Loeb L (1955). *Basic Processes of gaseous electronics*, University of California Press, Berkeley.

Lorikyan M.P. and Trofimchuk N.N. (1977). Particle detection by means of controllable secondary electron emission method. *Nucl Instr Meth* 140: 505–509.

Lubsandorzhiev B.K. (2006). On the history of photo-multiplier tube invention. *Nucl Instrum Meth A* 567: 236–238.

Maeda M. and Miyazoe Y. (1974). Progress in UV organic dye lasers. *IEEE J Quantum Elec* 10(9): 769–769.

Mardarskii O.I., Motorin O.V., Bologa M.K. and Kozhukhar I.A. (2000). Dispersion of liquid in an electric field of plane capacitor. *Proc IEEE 2000 Conf Elec Insul and Dirl Phen*, 292–295.

Martinelli R.U. and Fisher D.G. (1974). The application of semiconductors with negative electron affinity surfaces to electron emission devices. *Proc IEEE* 62: 1339–1360.

Massey H.S.W. (1950). *Negative ions*, 2nd edition, Cambridge University Press, Cambridge.

Masuda K., Aprile E., Ding H.L., Doke T., Gau S.S., Gaudreau M.P.J., Hitachi A., Ichinose H., Ishida N., Kase M., Kashiwagi T., Kikuchi J., Nakasugi T., Shibamura E., Sumorok K. and Takahashi T. (1991). Measurement of liquid xenon scintillation from heavy ions using a silicon photo-diode. *Nucl Instr Meth A* 309: 489–496.

McKinsey D.N., Brome C.R., Butterworth J.S., Glub R., Habicht K., Huffman P.R., Lamoreaux S.K., Mattoni C.E.H. and Doyle J.M. (1997). *Nucl Instrum Meth B* 132: 351–358.

McDaniel E.W. and Mason E.A. (1973). *The mobility and diffusion of ions in gases*, John Willey & Sons, New York–London–Sydney–Toronto.

Messous Y., Chambon B., Chazal V., De Jésus M., Drain D., Pastor C., de Bellefon A., Chapellier M., Chardin G., Gaillard-Lecanu E., Gerbier G., Giraud-Héraud Y., Lhote D., Mallet J., Mosca L., Perillo-Isaac M.-C., Tao C. and Yvon D. (1995). Calibration of a Ge crystal with nuclear recoils for the development of a dark matter detector. *Astropart Phys* 3: 361–366.

Meyer H.O. (2008). Dark Rate of a Photomultiplier at Cryogenic Temperatures. *e-print arXiv:0805.0771*.

Minday R.M., Schmidt W.F. and Davis H.T. (1971). Excess electrons in liquid hydrocarbons. *J Chem Phys* 54: 3112–3125.

Minehart R.C. and Milburn R.H. (1960). Scintillation-bubble chamber. *Rev Sci Instr* 31: 173–174.

Miyajima M., Mazuda K., Hitachi A., Doke T., Takahashi T., Konno S., Hamada T., Kubota S., Nakamoto A. and Shibamura E. (1976). Proportional counter filled with liquid xenon. *Nucl Instr Meth A* 134: 403–407.

Modinos A. (1984). Field, thermoionic, and secondary electron emission spectroscopy, Plenum Press, New York.

Monroe J. and Fisher P. (2007). Neutrino backgrounds to dark matter searches. *Phys Rev D* 76, 033007.

Moszynski M., Szawlowski M., Kapusta M. and Balcerzyk M. (2002). Large area avalanche photo-diodes in scintillation and X-rays detection. *Nucl Instr Meth A* 485: 504–521.

Ni K., Aprile E., Day D., Giboni K.L., Lopes J.A.M., Majewski P. and Yamashita M. (2005). Performance of a large area avalanche photo-diode in a liquid xenon ionization and scintillation chamber. *Nucl Instr Meth A* 551: 356–363.

Obodovskiy I. (2005). Solutes in rare gas liquids. In: *Electron Excitations in Liquefied Rare Gases*, Schmidt W.F. and Illenberger E. (eds), pp. 95–132, American Scientific Publishers.

Obodovsky I.M. and Pokachalov S.G. (1980). Electron capture by oxygen in condensed xenon; (1980). A new possibility to search for low-ionizing particles. In: *Experimental Methods of Nuclear Physics*, i.6, pp. 31–43, Atomizdat, Moscow (in Russian).

Ordonez C.E., Bolozdynya A.I., and Chang W. (1997). Doppler broadening of energy spectra in Compton scatter cameras. *1997 IEEE Nucl Sci Sym Med Imag Conf Rec* 2: 1361–1365.

Ovchinnikov B.M. and Parusov V.V. (2002). Shielding for low background experiments. *Astropart Phys* 17: 75–78.

Platzman R.L. (1961). Total ionization in gases by high-energy particles: an appraisal of our understanding. *Int J Appl Radiat Isot* 10: 116–127.

Policarpo A.P.L., Chepel V., Lopes M.I., Peskov V., Geltenbort P., Ferreira Marques R., Araújo H., Fraga F., Alves M.A., Fonte P., Lima E.P., Fraga M.M., Salete Leite M., Silander K., Onofre A. and Pinhão J.M. (1995). Observation of electron multiplication in liquid xenon with a microstrip plate. *Nucl Instr Meth A* 365: 568–571.

Radeka V. (1988). Low-noise Techniques in Detectors (1988). *Ann Rev Nucl Part Sci* 38: 217–277.

Ressel M.N. and Dean D.J. (1997). Spin-dependent neutralino-nucleus scattering for A = 127 nuclei. *Phys Rev C* 56: 535–546.

Rogers L., Clinthrone N., and Bolozdynya A. (2004). Compton cameras for nuclear medical imaging. *Emission Tomography: The Fundamentals of PET and SPECT*, Wernick M.N. and Aarsvold J.N. (eds), pp. 383–419, Academic Press, New York.

Rodionov B.U. (1969). Investigation of processes on tracks of ionizing particles in noble gases and liquids and a possibility of development of controllable track detector based on liquefied noble gases. PhD thesis, MEPhI, Moscow (in Russian).

Rodionov B.U. (1975). On project of emission chamber. *Experimental Methods of Nuclear Physics* 1: 36–45 (in Russian).

Rodionov B.U. (1979). Emission method of particles registration. In: *Elementary Particles*, pp. 64–76, 6th School of Physics, ITEP, Moscow (in Russian).

Rodionov B.U. (1987). Emission method of investigation of ionization processes in matter. Doctor of Phys and Math thesis, MEPhI, Moscow (in Russian).

Rossi B. and Staub H. (1949). *Ionization Chambers and Counters*, McGraw-Hill Book Company, Inc., New York.

Rubbia A. (2003). Experiments for CP violations: a giant liquid argon scintillation, Cerenkov and charge imaging experiment. *Proc II Int Workshop on Neutrinos in Venice, e-print hep-ph/0402110*.

Sadygov Z.Y., Zheleznykh I.M., Malakhov N.A., Jejer V.N. and Kirillova T.A. (1996). Avalanche semiconductor radiation detectors. *IEEE Trans Nucl Sci* 43: 1009–1013.

Samson J.A.R. (1967). *Technique of vacuum ultra-violet spectroscopy*, Pied Publications, Nebraska.

Sakai Y. (2005). Hot Electrons. In: *Electronic excitations in liquefied rare gases*, Schmidt W.F. and Illenberger E. (eds), pp. 275–294, American Scientific Publishers, California.

Seidel G.M., Lanou R.E. and Yao W. (2002). Rayleigh scattering in rare-gas liquids. *Nucl Instr Meth A* 489: 189–194.

Schmidt W.F. (1997). *Liquid state electronics of insulating liquids*, CRC Press LLC, New York.

Schmidt W.F., Hilt O., Illenberger E. and Khrapak A.G. (2005). The mobility of positive and negative ions in liquid xenon. *Rad Phys Chem* 74: 152–159.

Schnee R.W., Akerib D.S., Attisha M.J., Bailey C.N., Baudis L. *et al.* (2005). The SuperCDMS experiment. *Preprint arXiv:astro-ph/0502435 v1*, 21 February.

Schoepe W. and Dransfeld K. (1969). Extraction of electrons from quantised vortex lines. *Phys Lett A* 29: 165–166.

Schoepe W. and Rayfield G.W. (1973). Tunnelling from electronic bubble states in liquid helium through the liquid–vapor interface. *Phys Rev A* 7: 2111–2121.

Shibamura E., Hitachi A., Doke T., Takahashi T., Kubota S. and Miyajima M. (1975). Drift velocities of electrons, saturation characteristics of ionization and w-values for conversion electrons in liquid argon, liquid argon–gas mixtures and liquid xenon. *Nucl Instrum Meth* 131: 249–258.

Shikin V.B. and Leiderer P. (1981). To oscillations of the charged surface of liquid helium. *J Exp and Theor Phys* 81: 184–201 (in Russian).

Shimamori H. and Hatano Y. (1977). Thermal electron attachment to O_2 in the presence of various compounds as studied by a microwave cavity technique combined with pulse radiolysis. *Chem Phys* 21: 187–201.

Shutt T., Dahl C.E., Kwong J., Bolozdynya A. and Brusov P. (2006). Performance and Fundamental Processes at Low Energy in a Two-Phase Liquid Xenon Dark Matter Detector. *Nucl Instrum Meth A* 579 (2007): 451–453.

Shutt T., Dahl C.E., Kwong J., Bolozdynya A. and Brusov P. (2007). Performance and fundamental processes at low energy in a two-phase liquid xenon dark matter detector. *Nuclear Physics B: Proceedings Supplement* 173: 160–163.

Sidorov I.V. (1975). Cryogenic streamer chamber. *3d ITEP School* IV: 52–60 (in Russian).

Sinnok A.C. and Smith B.L. (1969). Refractive indices of the condensed inert gases *Phys Rev* 181: 1297–1307.

Solovov V.N., Chepel V., Lopes M.I., Hitachi A., Ferreira Marques R. and Policarpo A.J.P.L. (2004). Measurement of the refractive index and attenuation length of liquid xenon for its scintillation light. *Nucl Instrum Meth A* 516: 462–474.

Solovov V.N., Hitachi A., Chepel V., Lopes M.I., Ferreira Marques R. and Policarpo A.J.P.L. (2002). Detection of scintillation light of liquid xenon with a LAAPD. *Nucl Instrum Meth A* 488: 572–578.

Sowada U., Schmidt W.F., and Bakale G. (1977). The influence non-electronegative molecules on the mobility of excess electrons in liquefied rare gases and tetramethylsilane. *Can J Chem* 55: 1885–1889.

Spear W.E. and Le Comber P.G. (1977). Electronic transport properties. In: *Rare Gas Solids, vol. II*, Klein M.L. and Venables J.A. (eds), pp. 1118–1149, Academic Press, London.

Spence J., Castle J., and Webb J.H. (1948). Tracks of low ionizing particles in photographic emulsions. *Phys Rev* 74: 704–705.

Spicer W.E. (1977). Negative affinity III-V photo-cathodes: their physics and technology. *Appl Phys* 12: 115–130.

Spooner N.J.C., Davies G.J., Davies J.D., Pyle G.J., Bucknell T.D., Squier G.T.A., Lewin J.D. and Smith P.F. (1994). The scintillation efficiency of sodium and iodine recoils in a NaI(Tl) detector for dark matter searches. *Phys Lett B* 321: 156–160.

Stoletow A. (1890). On photo-electric currents in rarefied air. *J Phys* 9: 468–473.

Surko C.M. and Reif F. (1968). Investigation of a New Kind of Energetic Neutral Excitation in Superfluid Helium. *Phys Rev* 175: 229–241.

Surko C.M. and Reif F. (1968b). Evidence for a New Kind of Energetic Neutral Excitation in Superfluid Helium. *Phys Rev Lett* 20: 582–585.

Takahashi T., Konno S., Hamada T., Miyajima M., Kubota S., Nakamoto A., Hitachi A., Shibamura E. and Doke T. (1975). Average energy expended per ion pair in liquid xenon. *Phys Rev A* 12: 1771–1775.

Taylor G.I. and McEwan A.D. (1965). The stability of a horizontal fluid interface in a vertical electric field. *J Fluid Mech* 22: 1–15.

Trofimov V.N. (1996). R&D of large-sized cryogenic detectors in Dubna. *Nucl Instrum Meth A* 370: 168–170.

Troyanovky A.M., Volodin A.P. and Khaikin M.S. (1979). Electron localisation above the liquid hydrogen interface. *JETP Lett* 29: 65–68 (in Russian).

Veksler V., Groshev L. and Isaev B. (1949). *Ionization methods of investigation of radiations*, State Publishing Company of Technical-Theoretical Literature, Leningrad.

Volodin A.P. and Khaikin M.S. (1979). Ion 'geysers' on the surface of superfluid helium. *Pis'ma Zhurnal Eksperimental'noj i Teoreticheskoj Fiziki* 30: 608–610 (in Russian).

Volodin A.P., Khaikin M.S. and Edel'man V.S. (1976). Surface electron states above the helium film. *Pis'ma Zhurnal Eksperimental'noj i Teoreticheskoj Fiziki* 23: 524–527 (in Russian).

Voronova Ya. T., Guschin E.M., Kruglov A.A., Obodovsky I.M., Pokachalov S.G. and Shilov V.A. (1978). Installation for growing and testing crystals of noble gases. *Eksperimental'nye Metody Yadernoj Fiziki* 8: 63–69.

Walters A.J. (2003). Ion transport across the gas–liquid interface in xenon. *J Phys D Appl Phys* 36: 2743–2749.

Wright A.G. (1999). Absolute calibration of photo-multiplier based detectors – difficulties and uncertainties. *Nucl Instrum Meth A* 433: 507–512.

Yamashita M. (2003). Dark matter search experiment with double phase Xe detector. PhD thesis, Waseda University, Tokyo.

Yoshino K., Sowada U. and Schmidt W.F. (1976). Effect of molecular solutes on the electron drift velocity in liquid Ar, Kr, and Xe. *Phys Rev A* 14: 438–444.

YongLin J.U., Yan G.U., Dodd J., Galea R., Leltchouk M., Willis W., Rehak P. and Tcherniatine V. (2007). Detection of low energy solar neutrinos by two-phase cryogenic e-bubble detector. *Chin Sci Bull* 52: 3011–3015.

Zaklad H. (1971). A purification system for the removal of electronegative impurities from noble gases for noble liquid nuclear particle detectors. *Preprint UCRL-20690*, Lawrence Radiation laboratory of University of California, Berkley.

Zel'dovich Ya. B. and Khlopov Yu. M. (1981). To the possibility of studying nature of the neutrino mass in double-beta decay. *Pis'ma Zhurnal Eksperimental'noj i Teoreticheskoj Fiziki* 34: 148–153 (in Russian).

Zworykin V.K. and Ramberg E.G. (1949). *Photo-electricity and its application*, John Wiley & Sons, New York.

Index